W9-BRB-982

SHARKS
IN · QUESTION

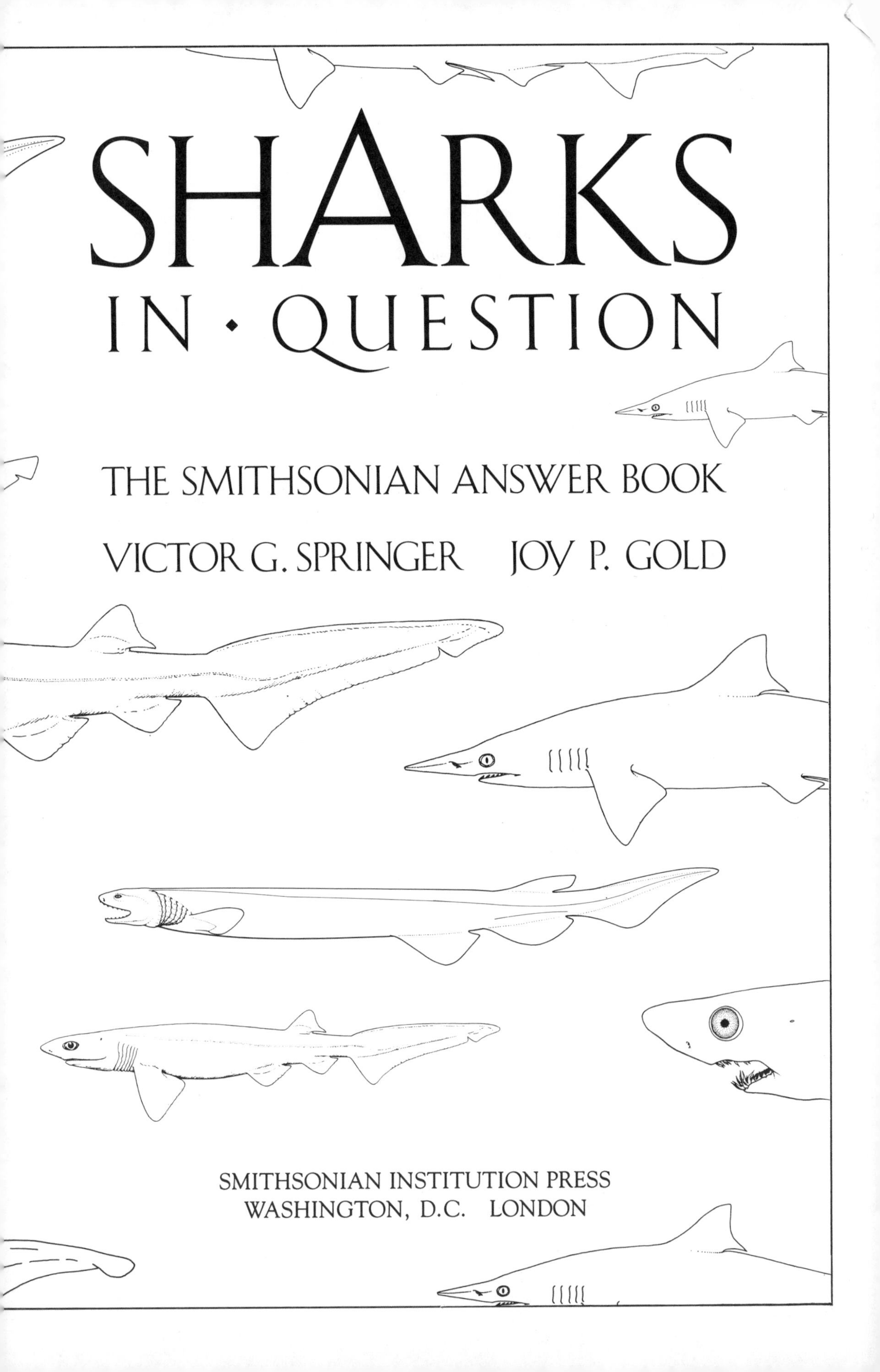

SHARKS IN · QUESTION

THE SMITHSONIAN ANSWER BOOK

VICTOR G. SPRINGER JOY P. GOLD

SMITHSONIAN INSTITUTION PRESS
WASHINGTON, D.C. LONDON

Royalties that would normally accrue to the authors from the sales of this volume will be deposited in the Smithsonian's Department of Vertebrate Zoology Discretionary Fund to be used for the increase and diffusion of knowledge about vertebrates, including sharks.

Designed by Lisa Buck Vann
Edited by Michelle K. Smith
Proofed by Leigh A. Alvarado

Library of Congress
Cataloging-in-Publication Data

Springer, Victor Gruschka, 1928–
Sharks in question: the Smithsonian answer book/
Victor G. Springer and Joy P. Gold.
p. cm.
Bibliography: p.
Includes indexes.
ISBN 0-87474-878-X.
ISBN 0-87474-877-1 (pbk.)
1. Sharks—Miscellanea.
I. Gold, Joy P.
II. Title.
QL638.9.S67 1989
597'.31—dc19 88–18185
CIP

Cover: The blue shark, *Prionace glauca.* (C. Scott Johnson)

Frontispiece: Whale shark, *Rhincodon typus,* Family Rhincodontidae, largest of all fishes, is a docile plankton feeder. (Tomoko Kimura)

Color printed in Hong Kong.
All other illustrations and text printed in the United States of America.
Manufactured in the United States of America.

10 9 8 7 6 5 4

99 98 97 96

∞ The paper used in this publication meets the minimum requirements of the American National Standard for Permanence of Paper for Printed Library Materials Z39.48–1984.

Contents

Preface

More than twelve years have passed since the first screening of *Jaws,* the Peter Benchley story and film adaptation of a maneating, shipwrecking great white shark. Since that event, sequels to the movie have been produced and the public has become preoccupied with fantasies of sharks as forces of evil. As a consequence of this heightened attention to the more notorious aspects of sharks, the Smithsonian has received thousands of letters and phone calls: some expressing strong opinions for or against sharks, some seeking verification of the more grisly scenes from the movies, some offering solutions for the "shark problem," and some requesting information about shark attacks, sharks in general, and the white shark in particular. The purpose of our book is to respond objectively to these concerns, to answer the questions, and to summarize what is known about the natural history, biology, and behavior of sharks as scientists understood them nearing the end of 1987.

Scientists are developing new information about sharks at a rapid pace, and the number of specialists, sometimes called elasmobranchologists, studying sharks and their close relatives (skates, rays, and chimaeras) appears to be increasing. Witness, in 1985 alone, the formation of the American Elasmobranch Society with a membership of 150 (now over 200), and the convening of four symposia on sharks (Baltimore, Maryland; Knoxville, Tennessee; Corvallis, Oregon; Tokyo, Japan) with international participation. As a result, our concepts on the biology and classification of sharks are in a state of constant change. Much of what we report will need revision or updating by the time it is published. There is much to learn, and the increased activity, in small part attributable to *Jaws,* makes these exciting times for scientists, fishermen, laymen, and perhaps even for sharks themselves.

This book was written for people with little prior knowledge of sharks, but with a curiosity and interest in finding out more about them. The first section of the book features the evolutionary history of sharks, their structure, how they swim, find food, eat, and reproduce, how they are classified, and the characteristics of the major groups (Orders) of sharks, with brief discussions of the natural history of representative species. The second section describes in more detail what is known about the life history and behavior of the "supersharks," the largest and/or most dangerous species. Following that, we discuss shark attacks, the Shark Attack File, shark repellents, and provide some suggestions for recognizing and minimizing the possibility of an attack. We conclude with recommendations on how to pursue a professional career as a shark expert.

The bibliography consists of two sections: (1) *General and Popular* and (2) *Further Readings.* The first section includes broad and comprehensive reviews of various aspects of sharks as well as easily read popular articles. The second section amplifies the text, providing the citations noted in further readings, which may also include some of the general and popular references.

Note about measurements and abbreviations. With few exceptions the metric system is used for measurements (including distances) and weights:

cm	=	centimeter	1 cm	=	approx. 0.4 inches
kg	=	kilogram	1 kg	=	approx. 2.2 pounds
km	=	kilometer	1 km	=	approx. 0.62 miles
m	=	meter	1 m	=	approx. 3.3 feet

m.y.a. = million years ago

Acknowledgments

Even in a small book such as this, we had substantive help from a large number of people in bringing this publication to completion. We here extend our sincere appreciation to all of them.

For generously providing photographs or drawings: George H. Burgess, Francis G. Carey, Ray Carthy, John Casey, Chip Clark, Eugenie Clark, David Doubilet, John F. Garrick, Perry W. Gilbert, Archie Gold, E. M. Grant, Samuel Gruber, Phillip C. Heemstra, C. Scott Johnson, Tomoko Kimura, Ad. J. Kalmijn, Sanford A. Moss, Amos Nachoum, Marta Nammack, Lisa J. Natanson, Gavin Naylor, Doug Perrine, Harold L. Pratt, Jr., Robert M. Pyle, John E. Randall, Jeffrey Siegel, Stewart Springer, T.F.H. Publications, Inc., National Marine Fisheries Service, Northeast Fisheries Center, Narragansett Laboratory, Narragansett, Rhode Island. Several of the line drawings were finely executed by T. Britt Griswold (Figures 1, 6b, 11, 13, 16, 27, 33, 35, 41, 45, 52).

For encouragement and help in important and innumerable ways: Warren E. Burgess, Steven Branstetter, John G. Casey, Guido Dingerkus, William N. Eschmeyer, Laurie Gold, Fran and Steven Goldsher, Nicholas Hotton III, Nancy Kohler, John E. McCosker, Harold L. Pratt, Jr., Maurice L. Rosenthal, B. L. Scronce, Jean Siegel, Gregg Skomal, Charles Stillwell, Doris Wey, Debra Wheaton, Mary Winters, and John P. Wourms.

Drafts of the manuscript were critically reviewed by professionals and laymen, who substantially improved its content: John F. Garrick, Robert H. Gibbs, Jr., Kazuhiro Nakaya, Archie Gold, W. Ron Heyer, Robert W. Purdy, Shirley S. Springer.

Last but not least, we express our appreciation to Theresa Slowik, science editor at Smithsonian Institution Press, for her efforts in propelling the manuscript through the publishing maze and a special thanks to Michelle Smith for her tireless editing and professional advice on ways to improve the text.

Royalties that would normally accrue to the authors from the sales of this volume will be deposited in the Smithsonian's Department of Vertebrate Zoology Discretionary Fund to be used for the increase and diffusion of knowledge about vertebrates, including sharks.

Introduction

Of all the marine organisms potentially dangerous to man, none are so universally feared and respected as are sharks. Perhaps, because we know so little about their life in the wild, they have taken on an aura of mystery, and even though far more people are fatally stung by wasps and bees, hit by cars, struck by lightning, or drowned every year than are attacked by sharks, we harbor a disproportionate amount of anxiety about shark attacks. On rare occasions, some sharks, such as the white, bull, and hammerhead, attack people, but far more frequently, we swim or work among these creatures unmolested, even unaware of their presence. Although sharks are not as great a threat to our well being as people seem to think, the power, mobility, invulnerability, and so-called unpredictability of sharks fascinate us.

Sharks are among the most successful predators in the sea, with few known enemies and an evolutionary history dating back more than 400 million years. They exhibit extreme diversity in form, behavior, and the habitats they occupy. They include species with awesome teeth or scarcely any teeth at all; giants over 15 meters long and dwarfs less than a quarter of a meter; species that live in the deep sea and luminesce and species that live in the shallows and are inconspicuous; species that swim across oceans and species that stay close to shore. And all these forms are endowed with exceptionally acute sensory systems and physical adaptations that have allowed them to persist into modern times.

In recent years, man, another successful predator, a terrestrial one, has expanded his recreational, exploratory, and food-gathering activities to invade, in continually increasing numbers, the shark's domain. Various government agencies and scientific and educational institutions explore shipwrecks and conduct underwater research. The U.S. Navy, in particular, is concerned with the protection of its equipment and the safety of its divers, pilots, and other personnel that work in, on, and over the sea. People are swarming to the ocean beaches to swim, boat, surf, and dive. Spearfishing, photography, and tropical reef exploration are popular. Sport fishing and shark fishing derbies attract enthusiastic numbers of people, and the number and size of commercial fisheries for sharks have increased in response to the public appetite for more and different sources of protein. How long sharks, especially the large, nearshore species, will be able to hold out against disruption of their habitat and destruction of their numbers is difficult to predict. Based on our experience with other marine species, it seems certain that some species of sharks will be depleted and, perhaps, extinguished by man's activities.

Whether the public is interested in sharks out of fear, competition, need of protection, or just curiosity, in writing this book the authors have been compelled to examine these remarkable animals and dispel some misconceptions concerning their behavior, even kindle some admiration for their life style. Knowledge leads to understanding and tolerance. As we learn more about sharks, perhaps we can accept them for what they are: other living creatures on this planet, eating, reproducing, protecting themselves, surviving, doing what they do best in their own milieu.

PART ONE

Shark Facts

WHAT ARE SHARKS?

Sharks are fishes and predators, armed with complex adaptations of teeth, jaws, skin, and sensory systems. They can see, hear, smell, navigate, and detect even hidden prey. They come in all sizes, ranging from enormous to tiny. They bear their young in more ways than do other groups of aquatic animals, and some of these modes of reproduction are more specialized than those of the bony fishes. They are masterful swimmers, and in the wild some may live to an old age of half a century. They can learn to perform tasks, even though their intelligence has been much maligned.

Sharks, like all other fishes, are vertebrates (animals with backbones). In the case of sharks and their relatives the "backbone," like the rest of their skeleton, is made of cartilage rather than bone. The material forming the skeleton thus distinguishes the two major groups of fishes that dominate the seas today, the bony fishes, or Osteichthyes, which constitute over 95 percent of all the species of fishes, and the cartilaginous fishes, or Chondrichthyes. The Chondrichthyes comprise two groups: the Elasmobranchii (sharks and rays, the latter including the skates) and the Holocephali (chimaeras).

How Are Sharks and Their Relatives Different from the Bony Fishes?

All sharks, as well as rays and chimaeras (Chondrichthyes) have a *cartilaginous* skeleton and this distinguishes them from the *bony* fishes. The skull and vertebrae of sharks may be heavily calcified, but with the possible exception of some bone at the base of their teeth, sharks are entirely boneless. There are many other differences between sharks and bony fishes, some of which will be discussed in other sections, but here we mention only two of the obvious ones. Sharks have five to seven external gill openings on each side of their head, whereas the bony fishes, like the chimaeras, have but one. And although sharks may have some short ribs, these never form a protective cage around the viscera as they do in the bony fishes.

How Do Sharks Differ from Skates, Rays, and Chimaeras?

The sharks and rays (figure 1) have at least five external gill openings on each side of the head, whereas, the chimaeras have but one. The chimaeras are a small group of fishes (probably fewer than 50 species) that live in the deep, cool waters of the Atlantic, Pacific, and Indian oceans (figure 2).

The differences between the sharks and rays (figure 3) are not absolute, as there are certain elasmobranchs, such as the sawfishes (figure 4) and guitarfishes (figure 5) (considered to be rays), and the sawsharks (figure 6) and angel sharks (figure 7), that have characteristics of both groups. Of the approximately 450 species of rays, most have evolved a depressed or flattened body form, especially in the head area. Their gill slits are located on the underside of the head, rather than on the sides of the head as in sharks. The edges of the pectoral fins of rays are attached to the side of the head over and in front of the gill openings, giving these fins the appearance of large, flat wings. In sharks, the edges of the pectoral fins are attached below or behind the gill openings. Some species of rays lack denticles (small tooth-like scales that cover the skin), but all sharks have them. There are also differences between the two groups in the way the jaw is suspended from the cranium (braincase).

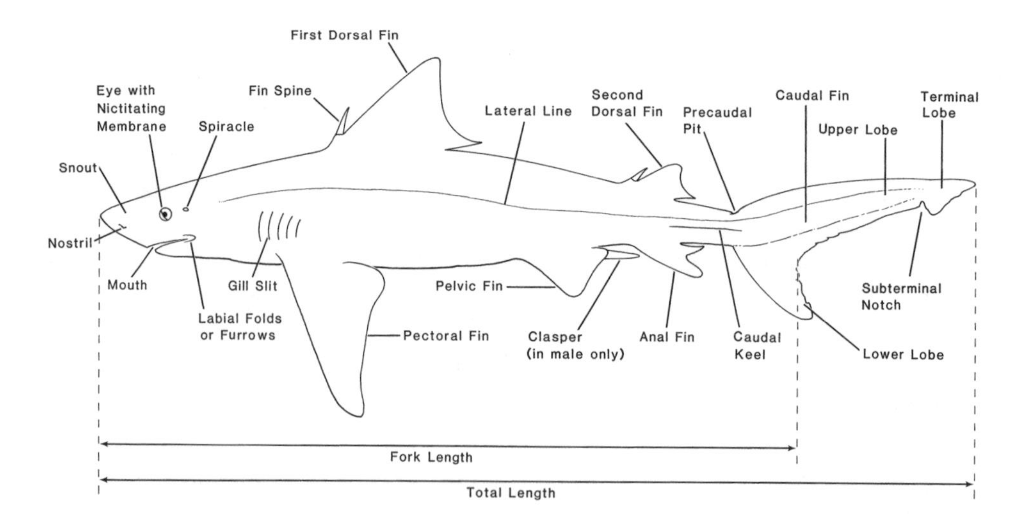

Figure 1. Composite shark showing principal parts and measurements.

Figure 2. A chimaera. (From Bigelow and Schroeder, 1953)

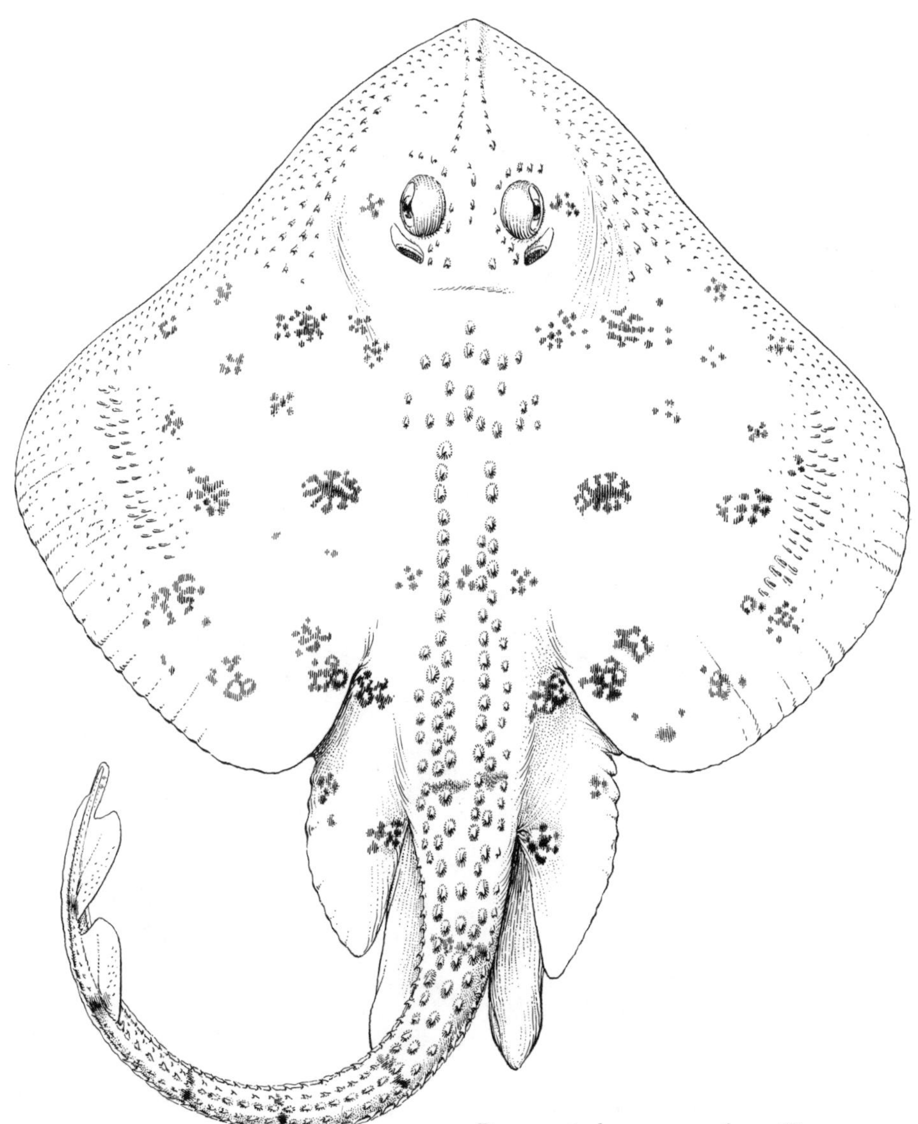

Figure 3. A skate, a type of ray. (From Bigelow and Schroeder, 1953)

Figure 4. A sawfish (note juvenile male claspers), a type of ray, dorsal (top), ventral (middle), and lateral aspects. (Smithsonian Institution)

Figure 5. A guitarfish (note male claspers), a type of ray, dorsal (below) and ventral aspects. (From Bigelow and Schroeder, 1953)

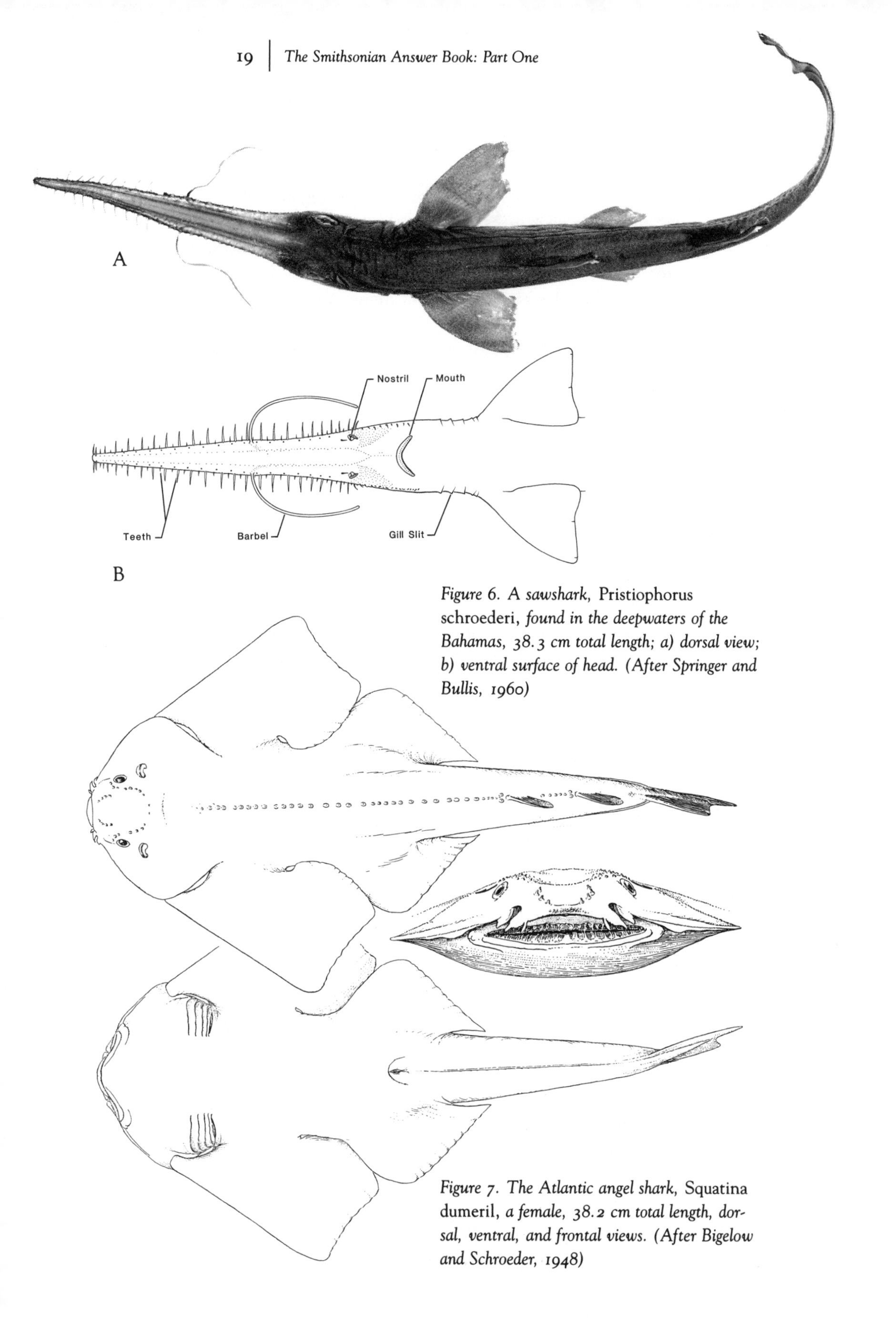

Figure 6. A sawshark, Pristiophorus schroederi, *found in the deepwaters of the Bahamas, 38.3 cm total length; a) dorsal view; b) ventral surface of head. (After Springer and Bullis, 1960)*

Figure 7. The Atlantic angel shark, Squatina dumeril, *a female, 38.2 cm total length, dorsal, ventral, and frontal views. (After Bigelow and Schroeder, 1948)*

WHAT IS CARTILAGE?

Cartilage, or gristle, is a kind of connective tissue strong enough to give support, and the same substance that gives shape to the human outer ear and nose. It is composed of single, paired, or clusters of cells, called chondrocytes, suspended in a matrix of protein. Cartilage may become calcified (contain deposits of calcium salts), or in the case of the bony fishes and higher vertebrates, invaded and replaced by bone cells (osteocytes). Often, the appearance of calcified cartilage and bone is very similar, and differences are apparent only after the tissues have been specially treated and examined with a microscope. According to some researchers, shark cartilage is distinct in composition from the cartilage of bony fishes and other higher vertebrates.

There is no reason to believe that having a cartilaginous skeleton is disadvantageous. On the contrary, cartilage is less dense and more elastic than bone, providing advantages of buoyancy and flexibility. This means that sharks do not have to work so hard to keep from sinking and their ability to maneuver is increased.

Sometimes differences in feeding preferences of sharks are reflected in the amount of calcification present in or on their jaws. For example, mollusk- and crustacean-eating sharks, such as nurse sharks, may have heavier jaws than sharks that feed primarily on fish. The heavier jaws are better able to withstand the pressures necessary to crush the hard parts of the prey.

Because cartilage and other soft parts decay rapidly, it is rare to find complete or even nearly complete fossil remains of sharks. This is one reason why the evolutionary relationships of sharks are so difficult to trace. Paleontologists have no way of knowing precisely what a shark living millions of years ago looked like merely by examining isolated teeth, spines, scales, and vertebrae, usually the only parts of a shark that remain in the fossil record.

Further readings
Caplan, A. J., 1984; Moss, M. L., 1977.

HOW ARE SHARKS ALIKE?

Sharks are remarkably diverse in size, behavior, and the habitats they occupy, yet they share numerous anatomical characteristics other than a cartilaginous skeleton. They all have a similar form, structure of the skin, teeth, jaws, reproductive organs, eyes, and intestinal valve.

Form

The streamlined shapes of torpedoes and submarines echo the efficient shape of sharks. Most species have a compact, fusiform body, two dorsal fins (no more than two and sometimes only one), a pair of pectoral and pelvic fins, a protruding snout with nostrils near the tip, and a crescent-shaped mouth underneath (figure 1). An anal fin, located between the pelvic fins and caudal fin (or tail), is present in many but not all sharks. Most sharks have five gill slits, or openings, but a few species have six or seven. Many sharks have an opening posterior to the eyes called a spiracle, through which water can be drawn and passed over the gills. The spiracle may be relatively large in bottom-dwelling species, in which it functions as an alternate pathway for respiration. It thus prevents the gills and mouth, through which sharks normally draw water for respiration, from clogging with mud and sand. The spiracle is an evolutionary remnant of the first gill slit of ancient fishes.

Skin

Shark skin is unique: tough, abrasive, and composed of tiny, hard, implanted tooth-like scales called placoid scales or denticles. *Denticle* means small tooth and indeed, it is from the denticle that the shark's real teeth have evolved (figure 8). The denticles are structurally similar to the teeth and have a hard outer layer of enamel (derived from the epidermis, the most superficial layer of the skin) overlying a core of dentine, nerve cells, and blood vessels. Like the teeth, the size and shape of the denticles may vary with the species and even slightly on different parts of the shark's body. In most sharks, the denticles form a complete coat over the body. Like the teeth, individual denticles do not increase in size with growth of the shark, but replacement denticles are larger than their predecessors. In the bramble sharks (*Echinorhinus,* figure 9) the denticles form large, prickly tubercles that are distributed irregularly over the body. Denticles are continually replaced throughout the shark's life.

Denticles are unlike the scales of bony fishes, which form without a contribution from the epidermis and usually lack core structures comparable to those in sharks. The scales of bony fishes generally are not shed, and do increase in size as the fish grows.

The skin of some sharks is so rough and abrasive that the victim of a shark attack may have large areas of his own skin scraped off by contact with the shark's body. Because of its texture, shark skin once was dried and made into a type of sandpaper called shagreen, which was used to polish wood. The skin is also cured, and after removal of the denticles, is used as leather, particularly for shoes.

Teeth

When one thinks "shark," one thinks teeth, and what remarkable teeth they are. It seems sharks have built their whole existence around their teeth. Shark teeth may be large or small, sharp or blunt, saw-like or knife-like, but above all they are endlessly replaceable. Shark teeth occur in multiple rows only in the jaws, in contrast to the bony fishes, which may have teeth on the gill arches, tongue, and roof of the mouth. Shark teeth, like denticles, consist of root, crown, and cusps or cusplets. The teeth are embedded in the gums of the jaws, rather than attached directly to the bone of the jaws as in other fishes.

Sharks have teeth of great variety (figure 10), and the number and shape of the

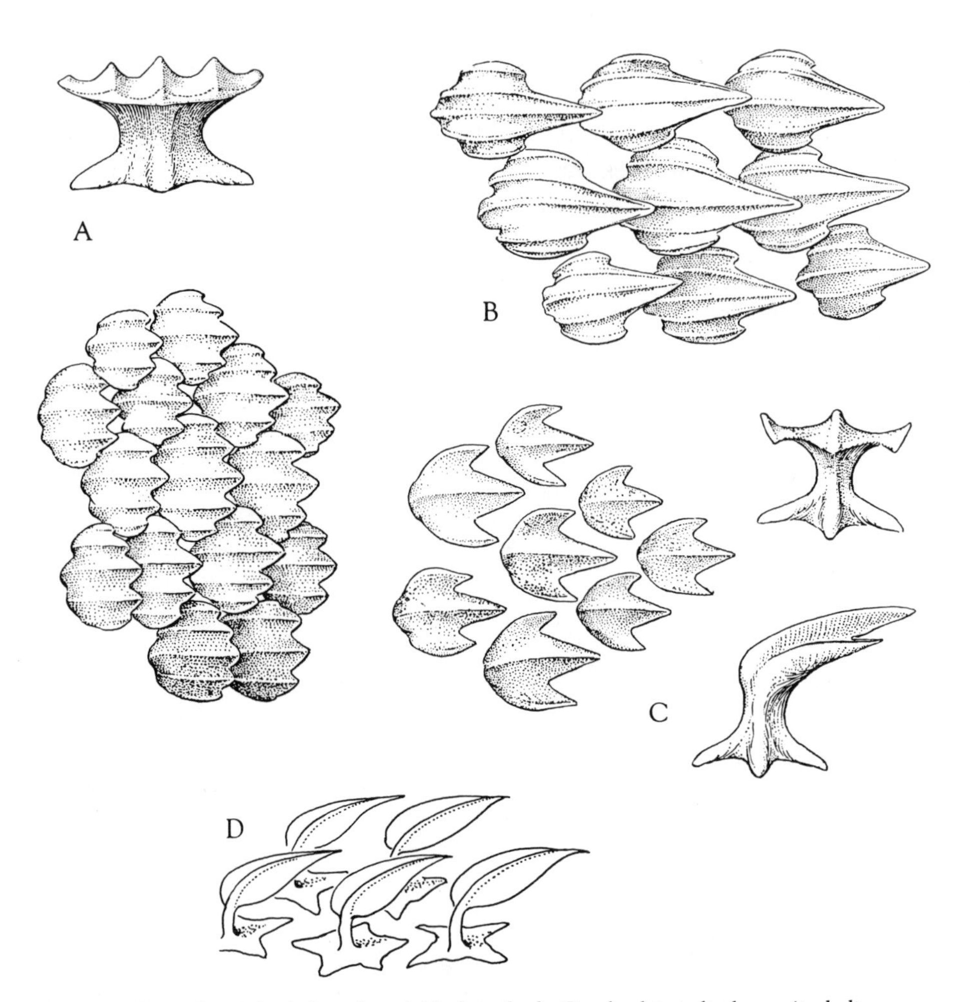

Figure 8. Examples of shark denticles: a) blacktip shark, Carcharhinus limbatus *(including frontal view of isolated denticle); b) chain dogfish,* Scyliorhinus retifer; *c) a cat shark,* Apristurus riveri *(including frontal and lateral views of isolated denticles); d) false cat shark,* Pseudotriakis microdon. *(From Bigelow and Schroeder, 1948)*

teeth are diagnostic characters frequently used to identify shark species, genera, and even families. Teeth may vary in shape and size, depending on which jaw they are in and their position in the jaw, front or rear. The largest teeth of the front row may be in the upper jaw, lower jaw, near the center of the jaw, somewhat posterior to the center of the jaw, or almost at the end of the jaw, or all the teeth may be similar in size and shape. In some species the teeth of juveniles differ in shape and size from those of adults, and in a few species, the shape and size of the teeth of males differ from those of females.

In most shark species only the front one or two rows of teeth are upright and functional, but a few species have as many as eight functional rows. The nonfunctional teeth are replacement teeth that lie with their tips or upper surfaces angled toward the rear of the mouth until the teeth are gradually erected and moved into position to fill the places of former functional teeth. The teeth increase in size throughout the shark's life, but not, however, because a tooth keeps growing throughout the shark's life. Only the replacement teeth that are still forming grow, but each replacement tooth is slightly larger than the functional tooth it replaced.

Replacement of teeth continues throughout the shark's life and for some species estimates are that an individual shark may shed as many as thirty-thousand teeth during its lifetime. Replacement may occur irregularly for single teeth or for whole rows at a time, depending on the species. White sharks and hammerheads appear to lose teeth irregularly, one by one. In contrast, in the spiny dogfish adjacent teeth overlap and form a continuous cutting edge, so it is more likely that in replacement an entire row moves upward and forward from its storage position. Tooth-replacement times are known for very few species of sharks. Captive lemon sharks about 1 m in length replace one row of functional teeth in their lower jaw every 8.2 days, those in the upper jaw every 7.8 days. Understandably, information on the replacement rates of the larger sharks is scarce. Records of teeth collected from a variety of captive sharks in oceanariums suggest these species replace their teeth every 6 to 12 months.

Because there is a relationship between the length of a shark and the height of its largest tooth, it is possible to draw a graph of tooth height against shark length for each species for which a good range of different size specimens is available (figure 11). It is then possible to reasonably estimate the length of an individual shark of one of these species based on the length of an isolated tooth, such as a fossil tooth, provided the tooth came from a species closely related to a living species and also came from the same position in the jaw of the fossil as the teeth upon which the graph was based. However, one cannot be absolutely certain about the length of the fossil shark, because there may be considerable variation in the

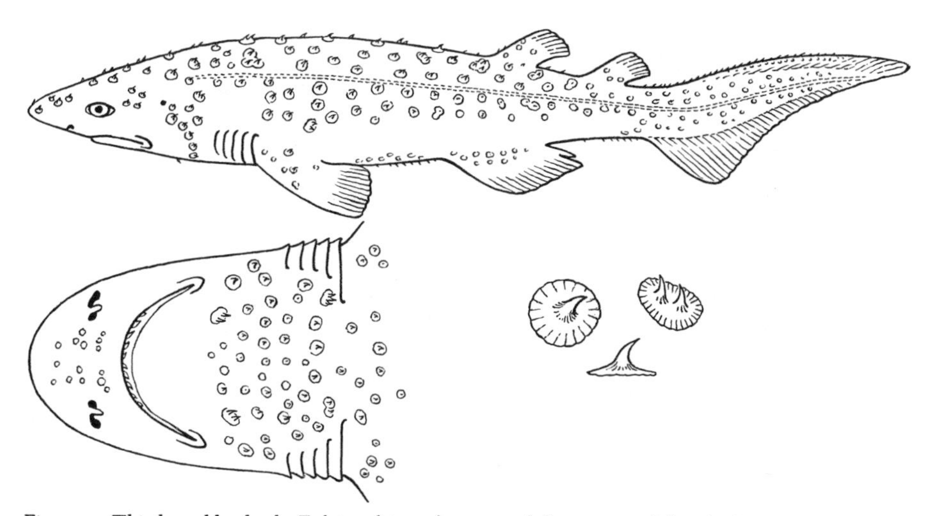

Figure 9. This bramble shark, Echinorhinus brucus, *of the western Atlantic, is a young individual about 92 cm total length (underside of head and isolated, enlarged denticles shown). In larger individuals spines on denticles are blunted or lost. (From Bigelow and Schroeder, 1948)*

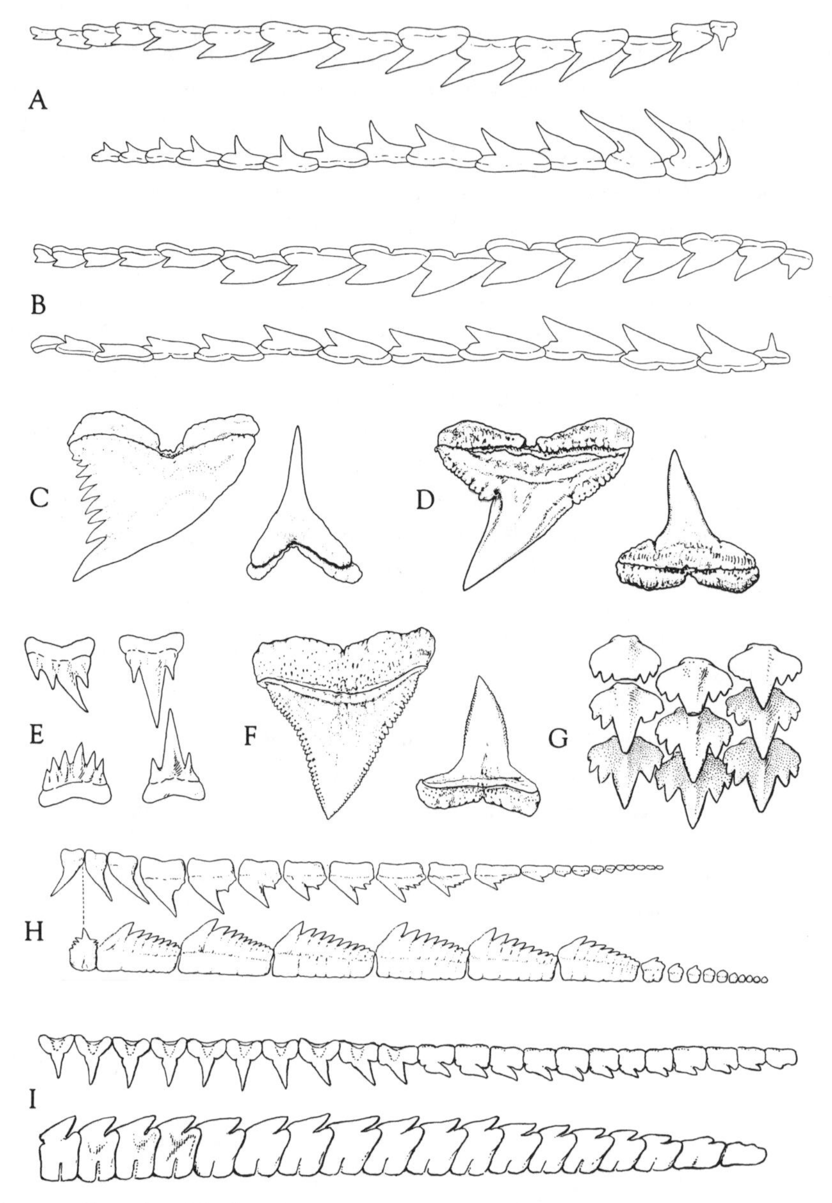

Figure 10. Examples of shark teeth (individual figures drawn to different scales): a) male; b) female: right side outer tooth rows of upper and lower jaws of spadenose shark, Scoliodon laticaudus *(From Springer, 1964); c) a weasel shark,* Hemigaleus microstoma, *selected right upper and lower teeth (Smithsonian Institution); d) night shark,* Carcharhinus signatus, *selected right upper and lower teeth (From Garrick, 1985); e) a cat shark,* Apristurus riveri, *selected upper and lower teeth from different parts of jaws (After Bigelow and Schroeder, 1948); f) oceanic whitetip,* Carcharhinus longimanus, *selected right upper and lower teeth (From Garrick, 1982); g) nurse shark,* Ginglymostoma cirratum, *sample of upper teeth. The lower teeth are similar (After Bigelow and Schroeder, 1948); h) sixgill shark,* Hexanchus griseus, *left side outer tooth rows of upper and lower jaws (From Bigelow and Schroeder, 1948); i) a dogfish,* Centroscymnus crepidater, *left side outer tooth rows of upper and lower jaws. (From Bigelow and Schroeder, 1957)*

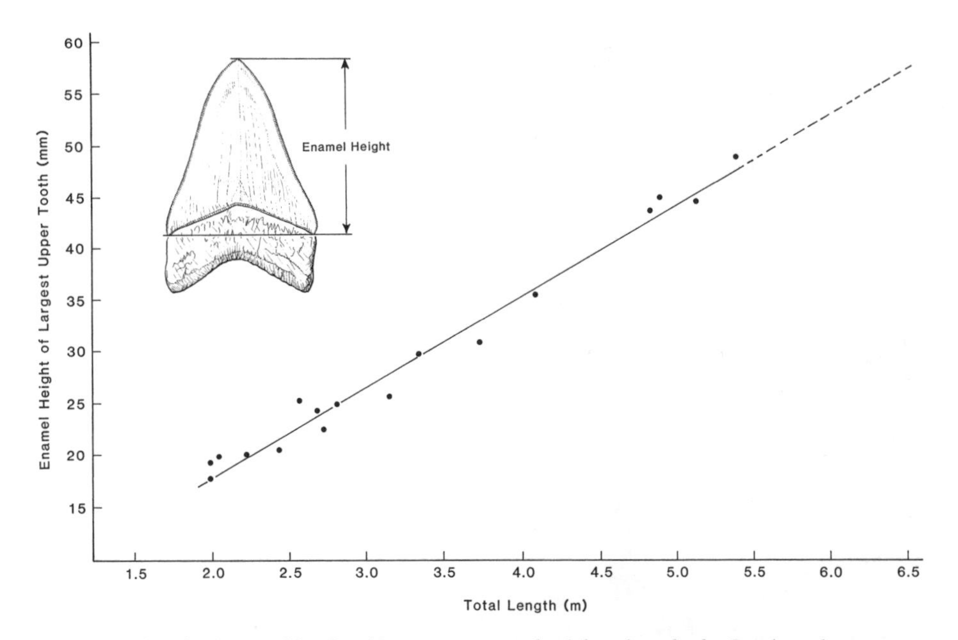

Figure 11. Graph of enamel height of largest upper tooth of the white shark, Carcharodon carcharias, *plotted against total length of shark. The solid line approximates the relationship between the two measurements. By extending the line (dashes) the size of a larger shark can be roughly estimated based only on the enamel height of its largest tooth. (Modified after Randall, 1973)*

size of the largest tooth in sharks of the same length and species. Some fossil shark teeth are much larger than any known for the same or closely related living species. To estimate the lengths of the ancient owners of these large teeth, one extends the line of the graph based on the related living species. The chances are small that the largest fossil tooth found for a particular species was the largest tooth from the proper position of a functional row, and estimates of shark length based on such a tooth are open to considerable error.

Terms such as "wicked," "vicious," and "razor sharp" have been used to describe the teeth of sharks. Indeed, those of some species are awesome structures for grasping, holding, cutting, and tearing prey. Some teeth, such as those found in the mako and sand tiger shark (plate 10d), are pointed and awl-shaped, well adapted for impaling fish or squid and holding them. Others, like those of the white shark, are efficient and effective saws: triangular, plate-like, with serrated edges, powerful enough to rip away chunks of flesh from animals larger than itself. Both types of teeth, triangular in the upper jaw and awl-shaped in the lower jaw, may be present in the same species of sharks (for example, several species of the requiem shark genus *Carcharhinus*). The smooth dogfishes and bullheads have rows of low, blunt teeth for crushing mollusks and crustaceans (figure 12). Other species of sharks have minute teeth. Those of the huge basking sharks and whale sharks are so small they scarcely conform with our idea of teeth, and they appear to play no role in feeding.

Figure 12. Lower jaw of the horn shark, Heterodontus francisci, *showing the distinctive crushing teeth. (Smithsonian Institution)*

Jaws

A broad underslung mouth armed with sharp teeth and powered with strong muscles that pull and push the jaws forward gives sharks a formidable feeding mechanism. The jaws of most modern sharks are loosely connected to the braincase and are short relative to the length of the head compared with those of many other predacious fishes. The loose attachment of the jaw allows sharks to widen their gape and extend the upper jaw so they can engulf large items of food, even whole animals. It also allows them to bite in any position, contrary to the popular belief that they must roll over to bite. But sharks cannot chew their food because they cannot move their jaws from side to side. Instead, after their teeth are firmly implanted in their prey, sharks may shake their heads vigorously until their teeth have sliced through the flesh. In this way they can cut chunks from prey too large to be eaten whole.

Do sharks have a tongue? Most sharks do, and they also have taste buds on the floor of the mouth and on the tongue. In sharks that have a tongue, the floor of the mouth cavity is raised up into a heavy fold that is supported internally by a special cartilage. All sharks have the cartilage, but not all sharks have a well-developed tongue. In the cookie-cutter shark the tongue is movable, and it is believed that as the shark presses its mouth against its prey, the tongue is retracted to form a vacuum. The suction of the vacuum might help the shark maintain a position against its prey.

Claspers

The males of all sharks (also skates, rays, and chimaeras) have claspers, which are modifications of the pelvic fins. Each male has two, one on each fin. The claspers are rod-like structures used during mating to transfer sperm to the female.

Mature males can be recognized by the relative size of their claspers. In some, if not all, sharks, the claspers appear to increase in length gradually as the shark grows until just before the onset of sexual maturity. At this time the claspers undergo a sudden spurt of growth and reach their maximum relative length, after which they continue to grow, but much more slowly.

Eyes

The eyes of sharks are generally similar in structure to those of other vertebrates, but with important differences. In most sharks and other fishes, the lens is almost spherical and constant in shape, whereas the typical land-dwelling vertebrate has a flattened or lentil-like lens whose shape can be changed to focus an image on the retina. Bony fishes focus by using a special muscle to pull the entire lens to or

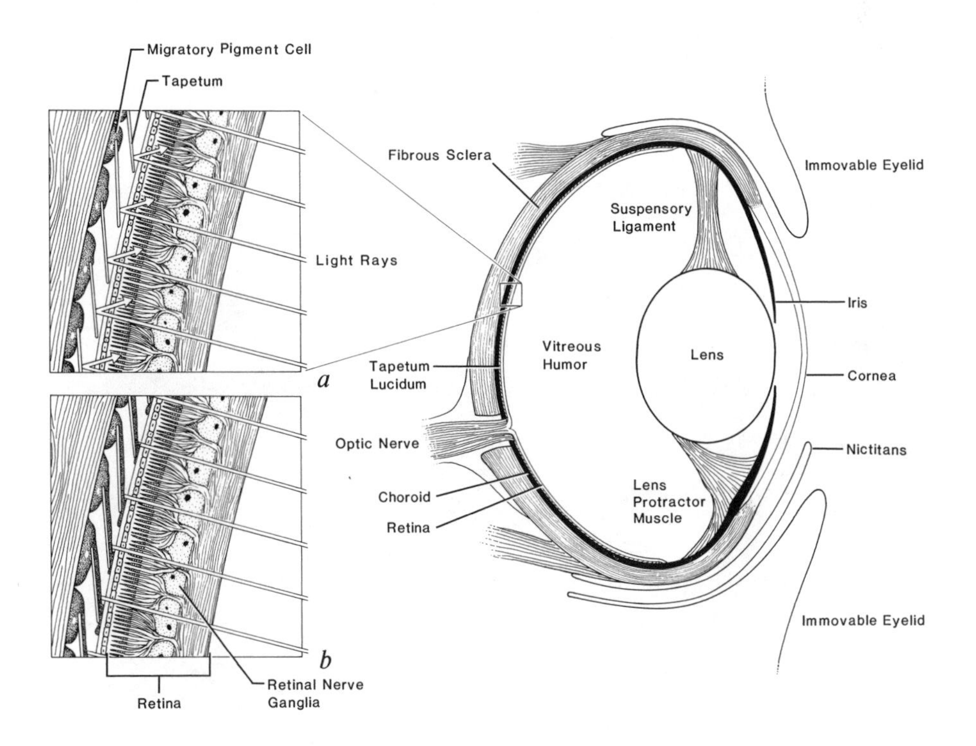

Figure 13. Structure of the elasmobranch eye (right). Enlarged sections through retina and tapetum lucidum: a) in dim light the migratory pigment cells are contracted and light passing through the retina strikes the tapetum lucidum and is reflected back through the retina, enhancing its sensitivity; b) in bright light the migratory pigment cells expand to cover tapetal plates, thus preventing reflection of the light. (Modified from Gilbert, 1984b)

away from the retina. It is not known whether, or how, sharks focus their eyes, because it has not been established whether the lens moves under normal circumstances (some movement has been inferred from experiments with anaesthetized sharks). If there is no movement or other accommodation, it would appear, in human terms, that sharks are moderately far-sighted. Although it remains to be determined how clearly sharks see, the eyes, rotated by prominent muscles, enable sharks to maintain a constant visual field and nearly panoramic vision, even when twisting and turning.

The pupils vary greatly in size and shape among sharks (figure 14). They may be round or slitlike, horizontally or vertically oriented. In some deep-sea species they do not dilate or constrict, but in many species they do. Changes in pupil size are a means of adapting to changes in light intensity: pupils dilate when light intensity is low and constrict when it is high. Maximum dilation and constriction are attained relatively slowly in most sharks that have been tested, but in some requiem sharks (*Carcharhinus*), the complete sequence takes less than a minute, which is much faster than occurs in humans. In the nurse shark (*Ginglymostoma cirratum*) dilation takes only 24 to 30 seconds, and constriction only 5 to 13 seconds.

Sharks have a remarkable adaptation, the tapetum lucidum (figure 13 a, b), which is present in some other vertebrates, such as the cat, which is known for its

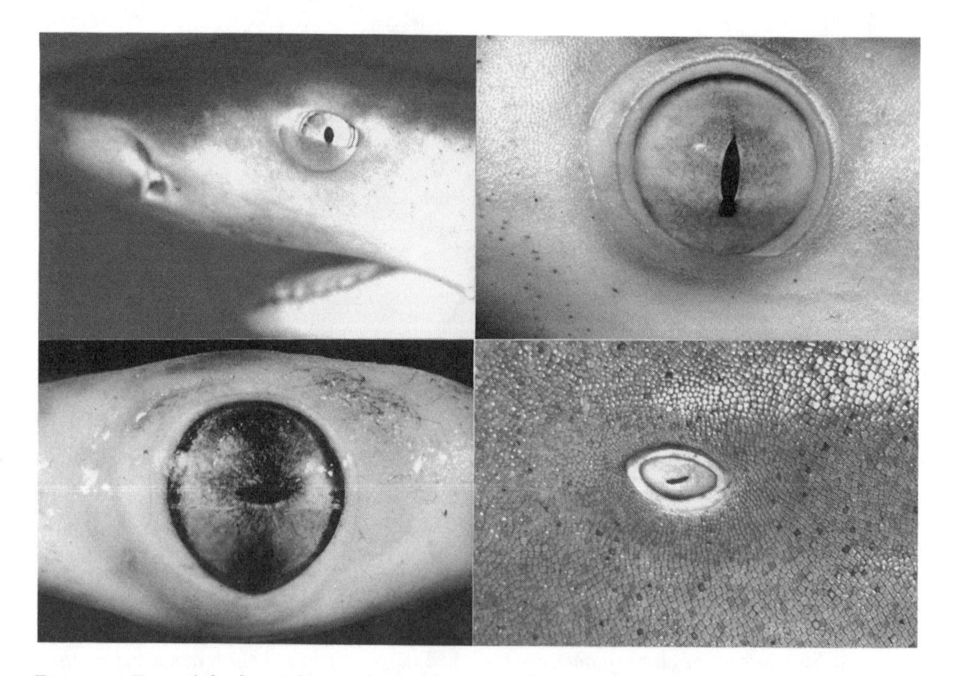

Figure 14. Eyes of sharks: a) lemon shark, Negaprion brevirostris; *note nictitating membrane, which is lacking in many sharks (E. Fisher); b) blacktip shark,* Carcharhinus limbatus *(Samuel H. Gruber); c) bonnethead shark,* Sphyrna tiburo *(Samuel H. Gruber); d) nurse shark,* Ginglymostoma cirratum. *(Gregg Skomal, National Marine Fisheries Service)*

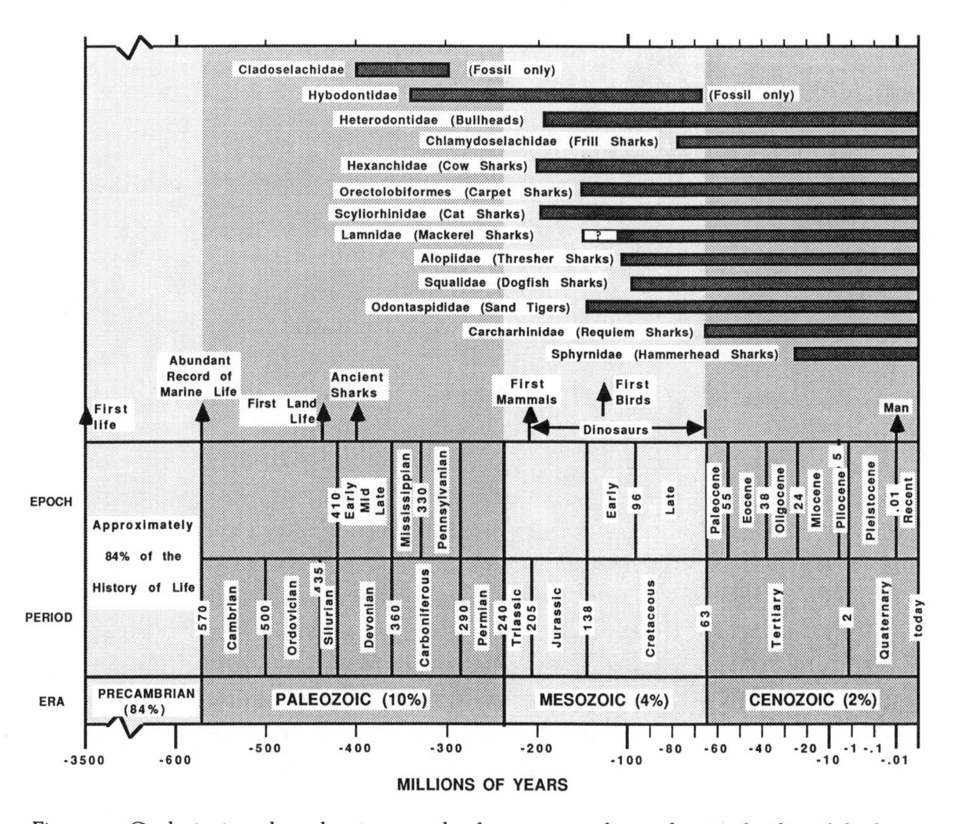

Figure 15. Geologic time chart showing records of occurrence of several major families of sharks (dark horizontal bars) and other groups of organisms throughout the history of life. (Compiled from Andrews et al., 1967; Cappetta, 1987; Geologic Names Committee, 1984; and Moy-Thomas, 1971)

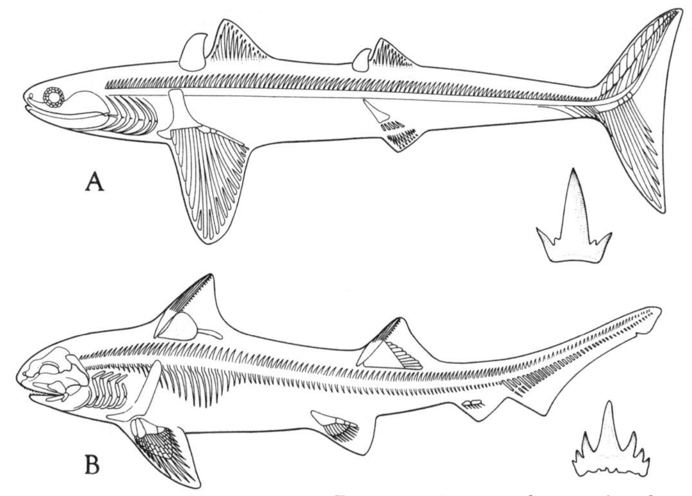

Figure 16. Ancient relatives of modern sharks: a) Cladoselache; *b)* Hybodus, *with representative tooth from each. (After Schaeffer and Williams, 1977)*

gape and protrusible jaw. Its teeth each consisted of a large central cusp with smaller lateral cusps, and were of a type paleontologists generally term "cladodont" (figure 18).

What was *Cladoselache*'s life style? It was probably an oceanic predator, feasting on smaller fish and sea animals that lacked shells, either swallowing them whole or tearing off pieces with its sharp teeth. Because it had large eyes, it probably detected its prey visually. *Cladoselache's* tuna-like tail suggests it was a strong, fast swimmer, but its broad-based pectoral and pelvic fins, unlike the narrow-based fins of modern sharks, may have limited its maneuverability. Large armored fishes dominated the Devonian seas, and it is not unlikely that *Cladoselache* served as a favored food item for these creatures. Although *Cladoselache* was probably an offshoot of the evolutionary line that gave rise to modern-day sharks (thus not a direct ancestor of living sharks), it probably closely resembled forms in that line.

By the end of the Carboniferous, 290 m.y.a., *Cladoselache* had disappeared, its place taken by more advanced sharks. In terms of the evolutionary line leading to modern sharks, the most important were a diverse assemblage of forms known as hybodonts (figure 16b). These fishes more closely resembled modern sharks in general appearance. They had narrow-based fins, which improved their maneuverability. An anal fin was present, as were claspers on the pelvic fins of males. Although some hybodonts retained the amphistylic jaw suspension of their earlier relatives, some species now had two kinds of teeth, sharp teeth in the front of the jaw and flat crushing ones in the rear. This permitted a more varied diet, which included mollusks and other shelled invertebrates. Hybodonts may have existed as early as the Upper Carboniferous (320 m.y.a.), flourished, but then, by the end of the Cretaceous (65 m.y.a.) became extinct.

One fossil genus of sharks, *Synechodus*, from the Late Cretaceous, was grouped with the hybodonts until recently. Dr. John G. Maisey, of the American Museum of Natural History, now believes that *Synechodus* is probably more closely related to modern sharks and rays than are the hybodonts. In the long run, this may only mean that the hybodonts, or some group of hybodonts, are ancestral to *Synechodus* and the modern elasmobranchs. Certainly, by the Cretaceous all modern shark families had developed and have changed little since that time.

The paleontology of modern sharks is evolving much faster than did the sharks themselves, and we can expect many changes in our ideas of shark history in the near future.

Perhaps no other shark evokes images of gigantic monsters or ferocious sea creatures than does *Carcharodon megalodon*, the extinct relative of the modern white shark. Fossil evidence for the genus *Carcharodon* dates back to the Paleocene (60–65 m.y.a.) but the largest teeth found belong to *C. megalodon* from the Pliocene, about 4.5 m.y.a. (plate 1). These teeth are 15 cm high, measured along a perpendicular from the upright tip to a flat surface at the base of the tooth. There is no doubt, from the size of these teeth, that their owner was the largest predatory fish swimming the seas of its time. Because the shape of the fossil teeth is almost identical to those of the modern white shark (*Carcharodon carcharias*), it is pre-

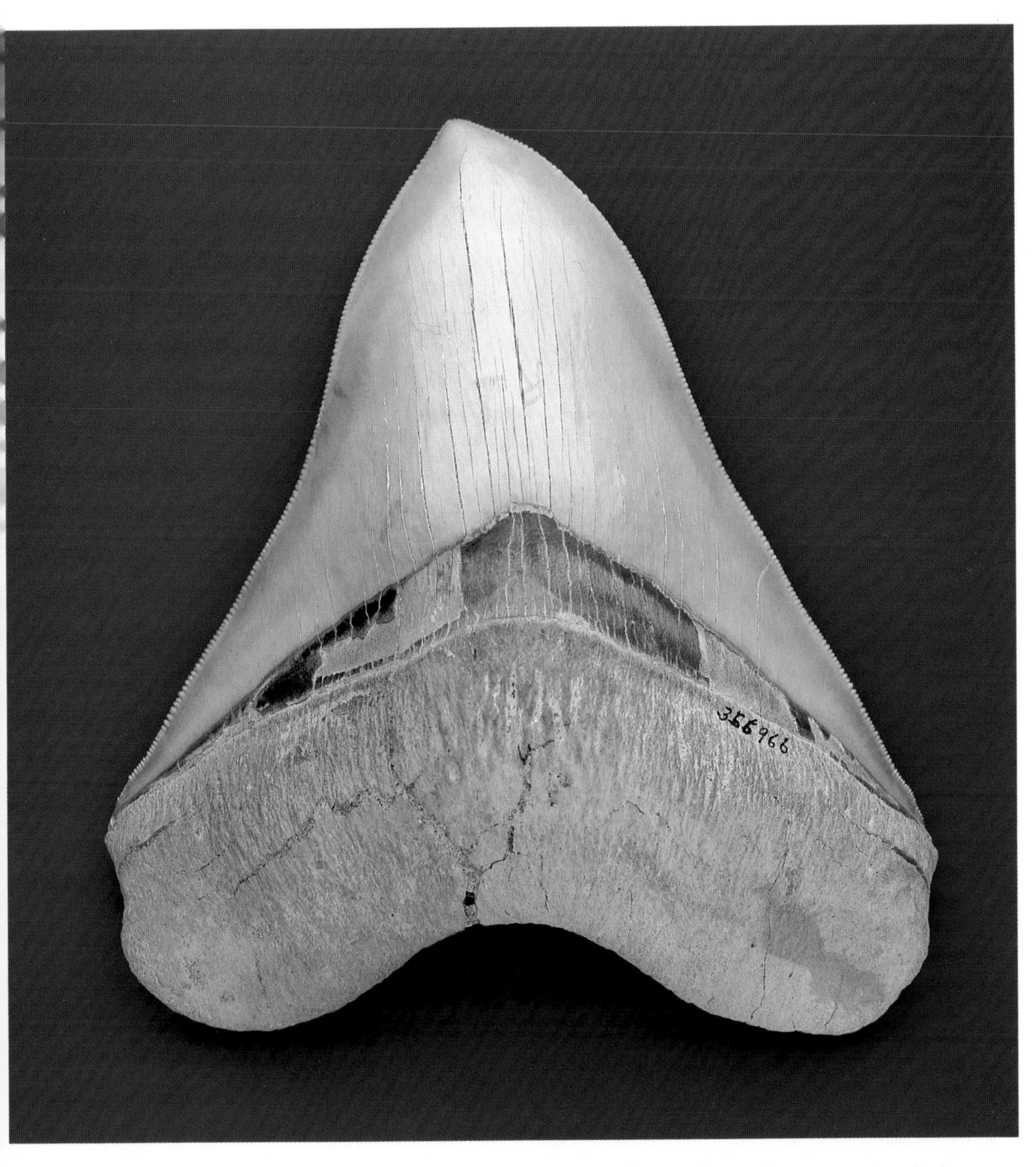

Plate 1. Tooth (actual size) of an extinct white shark, Carcharodon megalodon, *from the Pliocene, about 4.5 million years ago, of Lee Creek Mine, North Carolina. The owner is estimated to have been about 13 m (40 ft) long. (Chip Clark)*

Plate 2a. The brownbanded bambooshark, Chiloscyllium punctatum, *Family Hemiscyllidae, an attractive species restricted to the Indo-West Pacific. (Roger C. Steene, T.F.H. Publications, Inc.)*

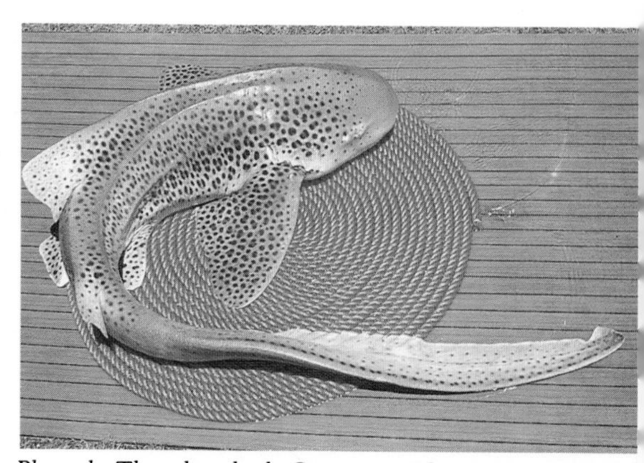

Plate 2b. The zebra shark, Stegostoma fasciatum, *Family Rhincodontidae, lives in the tropical Indo-West Pacific. (E. M. Grant)*

Plate 2c. The nurse shark, Ginglymostoma cirratum, *Family Rhincodontidae, found in the warm waters of the eastern Pacific and western and eastern Atlantic oceans. (National Marine Fisheries Service)*

Plate 2d. The speckled carpetshark, Hemiscyllium trispeculare, *Family Hemiscyllidae, is a shallow-dwelling species of northern Australian coral reefs. (Roger C. Steene, T.F.H. Publications, Inc.)*

Plate 2e. The swell shark, Cephaloscyllium ventriosum, *Family Scyliorhinidae, is a coastal inhabitant of the eastern Pacific. (Ken Lucas, T.F.H. Publications, Inc.)*

Plate 2f. The zebra bullhead shark, Heterodontus zebra, *Family Heterodontidae, is a beautiful inhabitant of the western Pacific. (K. H. Choo, T.F.H. Publications, Inc.)*

Plate 3a. A dwarf dogshark, Etmopterus perryi, *Family Squalidae, is one of the smallest shark species; this specimen is 182 mm total length. The dwarf dogshark is a deepwater inhabitant of the Caribbean Sea. (T. Britt Griswold)*

Plate 3b. The cigar shark, Squaliolus laticaudus, *Family Squalidae, is a deep-water inhabitant of the Atlantic, Indian, and western Pacific oceans. It is one of the smallest shark species. (Bruce H. Robison)*

Plate 3c. The Greenland shark, Somniosus microcephalus, *Family Squalidae, dwells in both shallow and deep waters of the northernmost Atlantic and Arctic oceans, also in the cold regions of the south Atlantic and Antarctic. The predatory Greenland shark attains a length of at least 6.4 m. (Roger Dow, National Marine Fisheries Service)*

Plate 4a. The reef shark, Carcharhinus perezi, *Family Carcharhinidae, lives around coral reefs in the tropical western Atlantic, and is noted for its habit of resting motionless on the bottom, as in this picture. (© 1987 Doug Perrine)*

Plate 4b. The gray reef shark, Carcharhinus amblyrhynchos, *is common around coral reefs of the Indian and western and central Pacific oceans. (John E. Randall)*

Plate 4c. The silvertip shark, Carcharhinus albimarginatus, *is seen frequently around the coral reefs of the tropical Indian and Pacific oceans. (Robert M. Pyle)*

Plate 4d. The oceanic whitetip shark, Carcharhinus longimanus, *found in all warm seas, usually occurs well offshore. (Ed Robinson, Marine Mammal Fund)*

Plate 5a. The whitetip reef shark, Triaenodon obesus, *Family Carcharhinidae, is a harmless reef inhabitant of the tropical Indian and Pacific oceans. A remora (or shark-sucker) has attached itself to the shark; the remora feeds on parasitic copepods on the shark's skin, mouth, and gills. (John E. Randall)*

Plate 5b. The blue shark, Prionace glauca, *Family Carcharhinidae, is a pelagic species, abundant in all temperate and tropical seas. (C. Scott Johnson)*

Plate 5c. The tiger shark, Galeocerdo cuvier, *Family Carcharhinidae, is present in all warm seas and has been implicated in many shark attacks. This young individual exhibits spots, which change to bars and fade with age. A white tag is visible near the first dorsal fin. (© 1987 Doug Perrine)*

Plate 5d. The lemon shark, Negaprion brevirostris, *Family Carcharhinidae, lives along tropical coasts of the Western Hemisphere and eastern Atlantic. (© 1987 Doug Perrine)*

Plate 6a. This white shark, Carcharodon carcharias, *Family Lamnidae, is a 4.8 m, 1,086 kg male. (H. Wes Pratt, Jr., National Marine Fisheries Service)*

Plate 6b. The shortfin mako, Isurus oxyrinchus, *Family Lamnidae. (Charles Stillwell, National Marine Fisheries Service)*

Plate 6c. The porbeagle, Lamna nasus, *Family Lamnidae, a coastal and oceanic species that is found in the cool temperate seas well north and south of the equator (bi-temperate distribution), but is notably absent from the north Pacific. This individual was approximately 2 m in total length and weighed about 82 kg. (Gregg Skomal, National Marine Fisheries Service)*

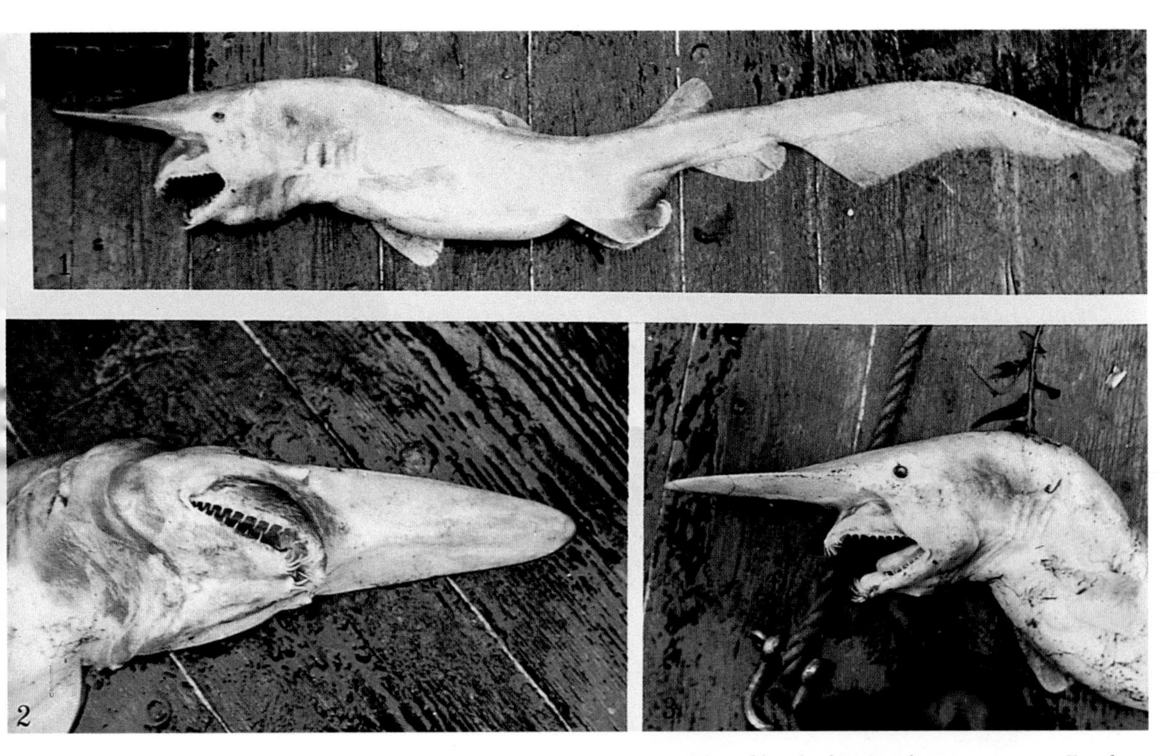

Plate 7a. The goblin shark, Mitsukurina owstoni, *Family* Mitsukurinidae*, is a rare, deep-dwelling shelf species found off most continental coasts (jaws protruded as if preparing to bite). (From Uyeno et al., 1976)*

*Plate 7b. Copepod parasites (*Pandarus *species) attached to the pectoral fin of a blue shark. (Alan Lintala, National Marine Fisheries Service)*

Plate 7c. Spiral valve of the white shark sliced open to show the "threads" of the valve, which increase the absorptive surface of the intestine and slow down the passage of food. (Jack Casey, National Marine Fisheries Service)

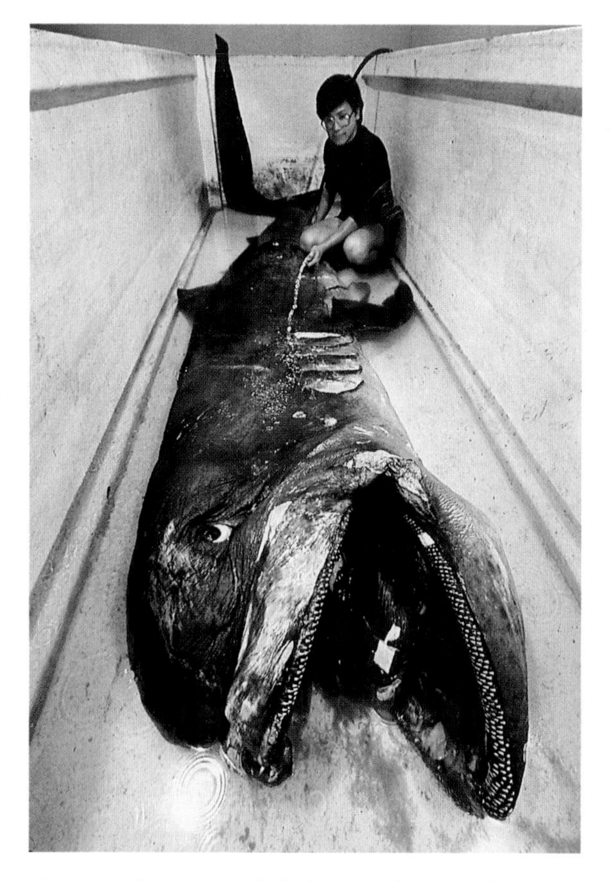

Plate 8a. The megamouth shark, Megachasma pelagios, *Family* Megachasmidae, *is known only from two specimens caught off the Hawaiian Islands and California. (David Doubilet, © 1981 National Geographic Society)*

Plate 8b. Mouth of the megamouth shark. Note the needle-like teeth on gums. Megamouth feeds on plankton. (Los Angeles County Museum)

Plate 8c. The bonnethead shark, Sphyrna tiburo, *Family* Sphyrnidae, *lives in warm coastal waters of the eastern Pacific and western Atlantic. The bonnethead is the smallest species of hammerhead, attaining a maximum total length of only 1.5 m. (Jack Casey, National Marine Fisheries Service)*

Plate 9a. The great hammerhead shark, Sphyrna mokarran, *Family Sphyrnidae, occurs worldwide in warm seas, attains a length of about 6 m. This male is about 3 m total length. (Guido Dingerkus)*

Plate 9b. This thresher shark, Alopias vulpinus, *Family Alopiidae, is approximately 3 m total length. (H. Wes Pratt, Jr., National Marine Fisheries Service)*

Plate 9c. The bigeye thresher, Alopias superciliosus. *(H. Wes Pratt, Jr., National Marine Fisheries Service)*

Plate 9d. The blacktip reef shark, Carcharhinus melanopterus, *Family Carcharhinidae, is a small, but potentially dangerous, shallow-dwelling inhabitant of the Indo-West and central Pacific oceans. (Walter Deas)*

Plate 10a. The tiger shark, Galeocerdo cuvier, *has distinctive teeth, and has been implicated in numerous attacks on man. (J. E. Randall)*

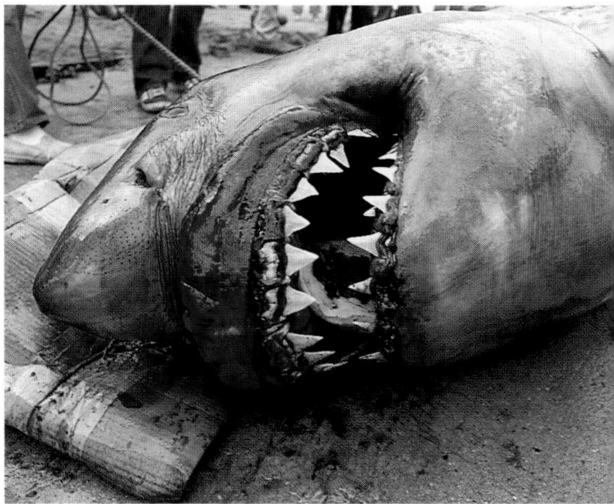

Plate 10b. The white shark, Carcharodon carcharias; *note the triangular, serrated teeth. (Alan Lintala, National Marine Fisheries Service)*

Plate 10c. The shortfin mako, Isurus oxyrinchus; *note the sensory pores, called ampullae of Lorenzini, on the underside of the snout. (Alan Lintala, National Marine Fisheries Service)*

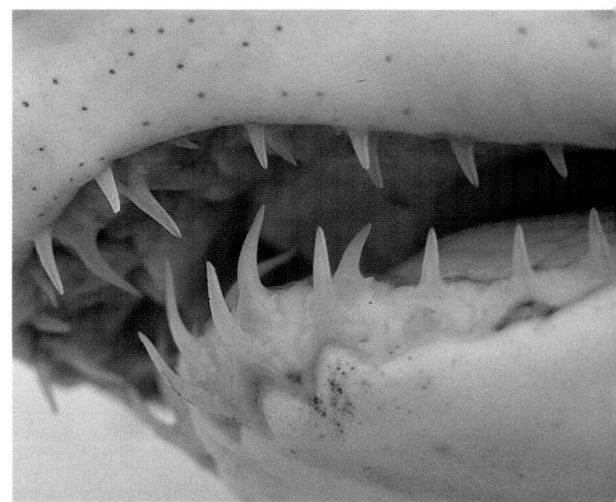

Plate 10d. The sand tiger shark, Eugomphodus taurus, *has sinuous, dagger-like teeth with a small cusp on either side of the base of the main cusp. (Gregg Skomal, National Marine Fisheries Service)*

Plate 11a. A blue shark accidentally caught in a net full of squid. (National Marine Fisheries Service)

Plate 11b. The large toothless gape of the basking shark, Cetorhinus maximus, *a plankton feeder, permits the shark to strain large volumes of water. The gill rakers, or strainers, are shed in the winter, and are not present in this animal (gill filaments are visible toward the bottom of the internal gill opening just behind the prop. (H. Wes Pratt, Jr., National Marine Fisheries Service)*

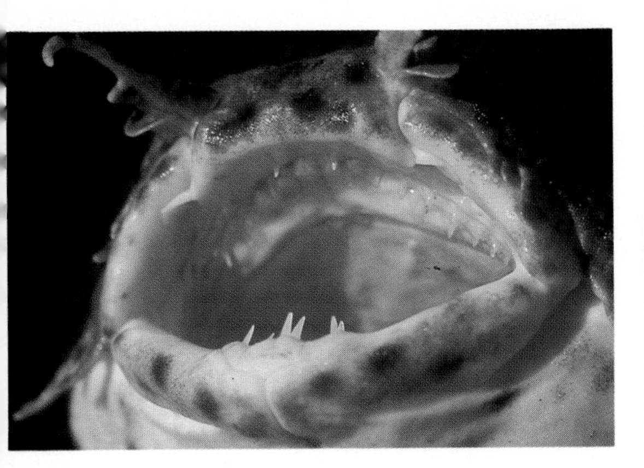

Plate 11c. A wobbegong, Orectolobus ornatus, *found in the shallow waters of Australia, New Guinea, and Japan, has distinctive grooves between the nostrils and upper jaw, a cleft chin, nasal barbels, and fleshy lobes, as well as needle-like teeth. (E. M. Grant)*

Plate 11d. A large white shark exhibits its enormous gape. (Amos Nachoum)

Plate 12a. Blue sharks, Prionace glauca, *are frequently attracted to fishing boats. (C. Scott Johnson)*

Plate 12b. A wobbegong, Orectolobus wardi, *lives on the continental shelf of northern Australia (note the large spiracle ventroposterior to the eye and the fleshy lobes on the side of the head; the nasal barbels probably have a sensory function). (Roger C. Steene, T.F.H. Publications, Inc.)*

Plate 12c. The Atlantic angel shark, Squatina dumeril, *Family Squatinidae, found in the warm waters of the western north Atlantic. This specimen bears a white dart tag. (National Marine Fisheries Service)*

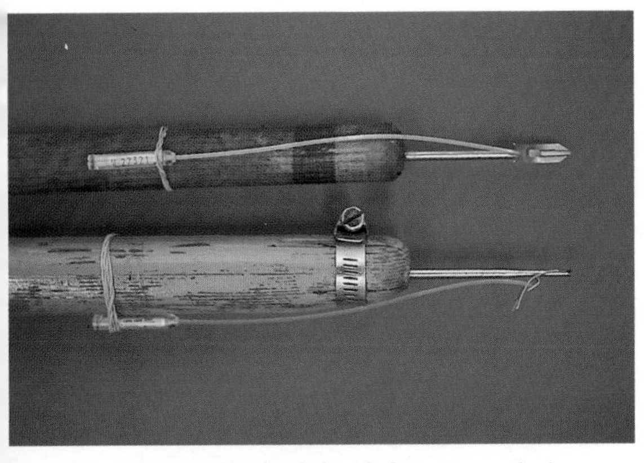

Plate 13a. Tagging poles loaded with dart tags ready for application. (H. Wes Pratt, Jr., National Marine Fisheries Service)

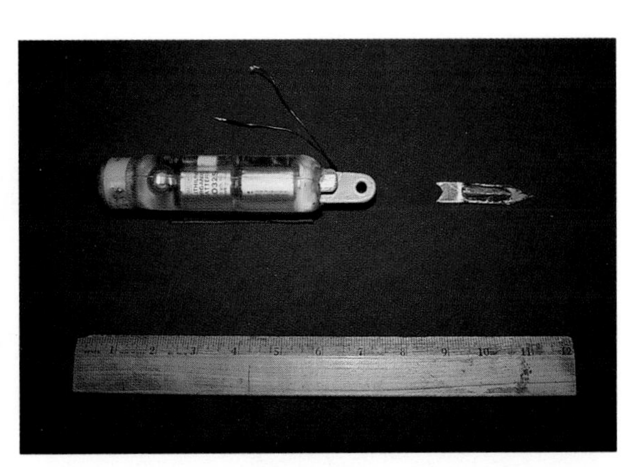

Plate 13b. A type of sonic tag used on sharks (H. Wes Pratt, Jr., National Marine Fisheries Service)

Plate 13c. Tiger shark with newly applied dart tag. The fishing line will be cut and the shark will cast off the hook later. (Gregg Skomal, National Marine Fisheries Service)

Plate 13d. Blue shark with recently applied orange sonic tag. (Francis G. Carey)

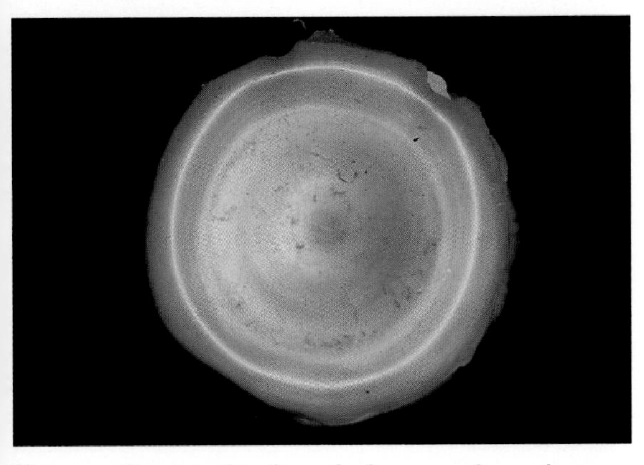

Plate 13e. Cross-sectioned vertebral centrum from a lemon shark treated with tetracycline: under ultraviolet light tetracycline fluoresces, showing growth rings (annuli) formed before and after marking in stained vertebrae. (S. H. Gruber)

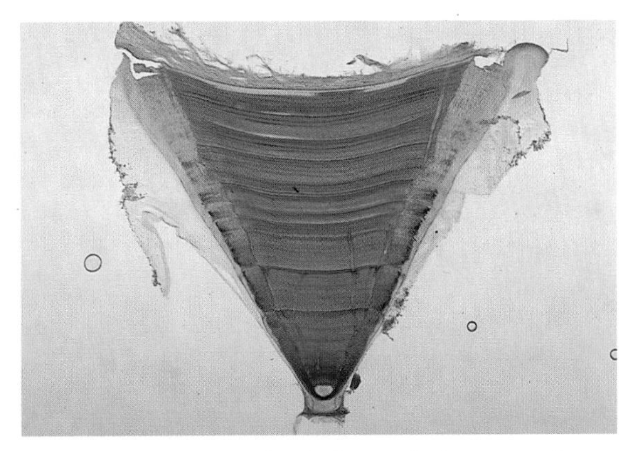

Plate 13f. Long section of half of a vertebral centrum of a sandbar shark stained to show the annuli. There are 10 annuli with some growth after the last annulus. (H. Wes Pratt, Jr., National Marine Fisheries Service)

Plate 14a. A spiny dogfish, Squalus acanthias, *with two "candles," one in each uterus. The upper candle, with three embryos, has been removed from its uterus; the lower candle is seen through the wall of its uterus, which is continuous with the dark slender oviduct that extends between buff-colored lobes of the liver. To the left of the oviduct is the whitish stomach, which is continuous with the posteriorly extending intestine. (Marta Nammack)*

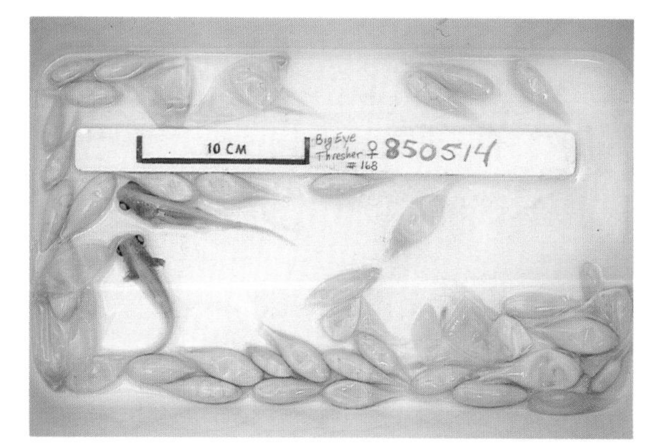

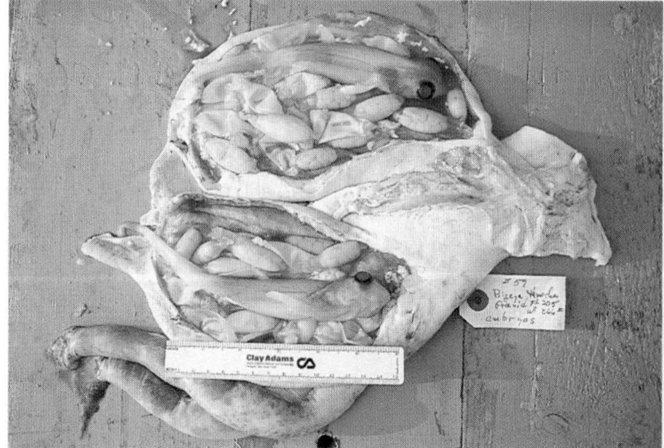

Plate 14 b and c. Embryos of the bigeye thresher, Alopias superciliosus, *in early and more advanced (with uteri) stages. The embryos feed on the packets of unfertilized eggs. Only one or two embryos survive in each uterus. (Gregg Skomal and Charles Stillwell, National Marine Fisheries Service)*

Plate 15a. A litter of near-term embryos of the scalloped hammerhead, showing umbilical cords attached to the uterus (Charles Stillwell, National Marine Fisheries Service)

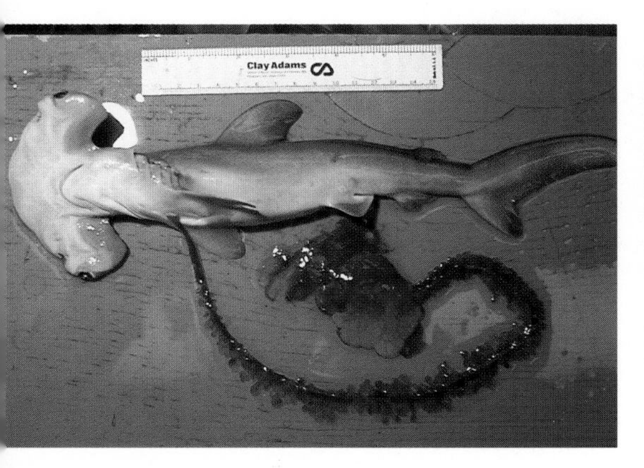

Plate 15b. Isolated embryo of the scalloped hammerhead, showing placenta and umbilical cord. (Alan Lintala, National Marine Fisheries Service)

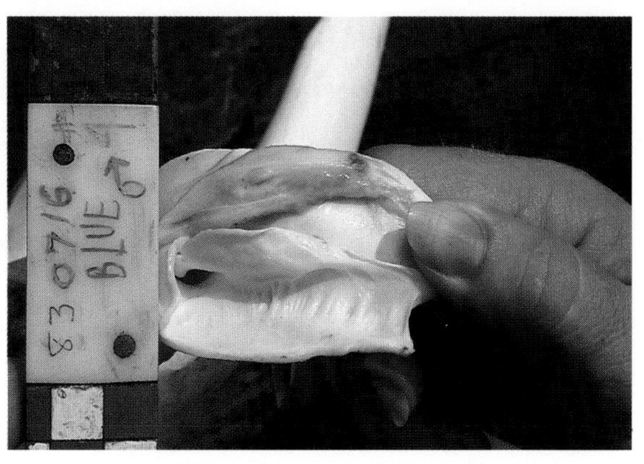

Plate 15c. Clasper of a male blue shark unrolled to show deep grooves and entry (dark hole) to siphon sac. (H. Wes Pratt, Jr., National Marine Fisheries Service)

Plate 16. Reconstructed jaws of Carcharodon megalodon, *a 13-meter shark that lived in the western Atlantic 4.5 million years ago. These jaws are on exhibit at the Smithsonian's National Museum of Natural History. Frequently published photographs of an earlier reconstruction showing jaws much larger than these are greatly in error. The earlier reconstruction was mistakenly made using only large teeth, all of about equal size. (Chip Clark)*

sumed that the two species are related, possibly evolving in the distant past from a common ancestor, or that the present-day white shark may be a direct descendant of the species for which we have only fossil teeth.

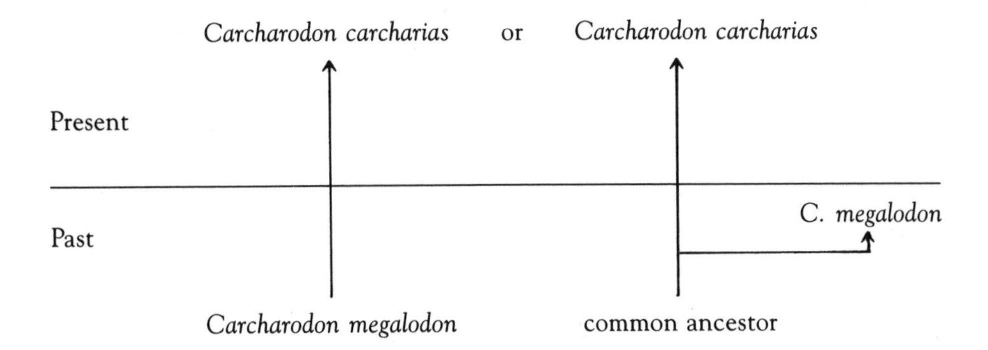

Carcharodon megalodon teeth were probably known to naturalists long before 1835, when the species was first assigned its scientific name. The size of the beast has been a subject for discussion ever since. By comparing the fossil teeth with the jaws and teeth of the modern white shark, *C. carcharias*, length estimates ranging up to 30 m (about 100 ft) have been made. However, many of the calculations used to determine reconstructions of the fossil jaws and estimates of length are now

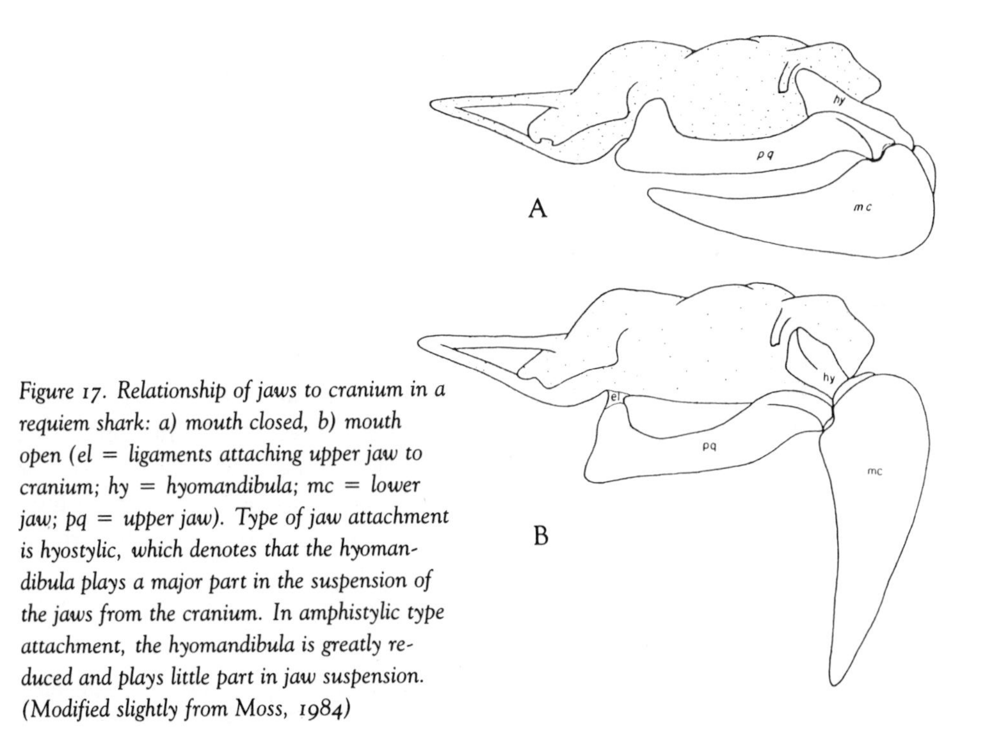

Figure 17. Relationship of jaws to cranium in a requiem shark: a) mouth closed, b) mouth open (el = ligaments attaching upper jaw to cranium; hy = hyomandibula; mc = lower jaw; pq = upper jaw). Type of jaw attachment is hyostylic, which denotes that the hyomandibula plays a major part in the suspension of the jaws from the cranium. In amphistylic type attachment, the hyomandibula is greatly reduced and plays little part in jaw suspension. (Modified slightly from Moss, 1984)

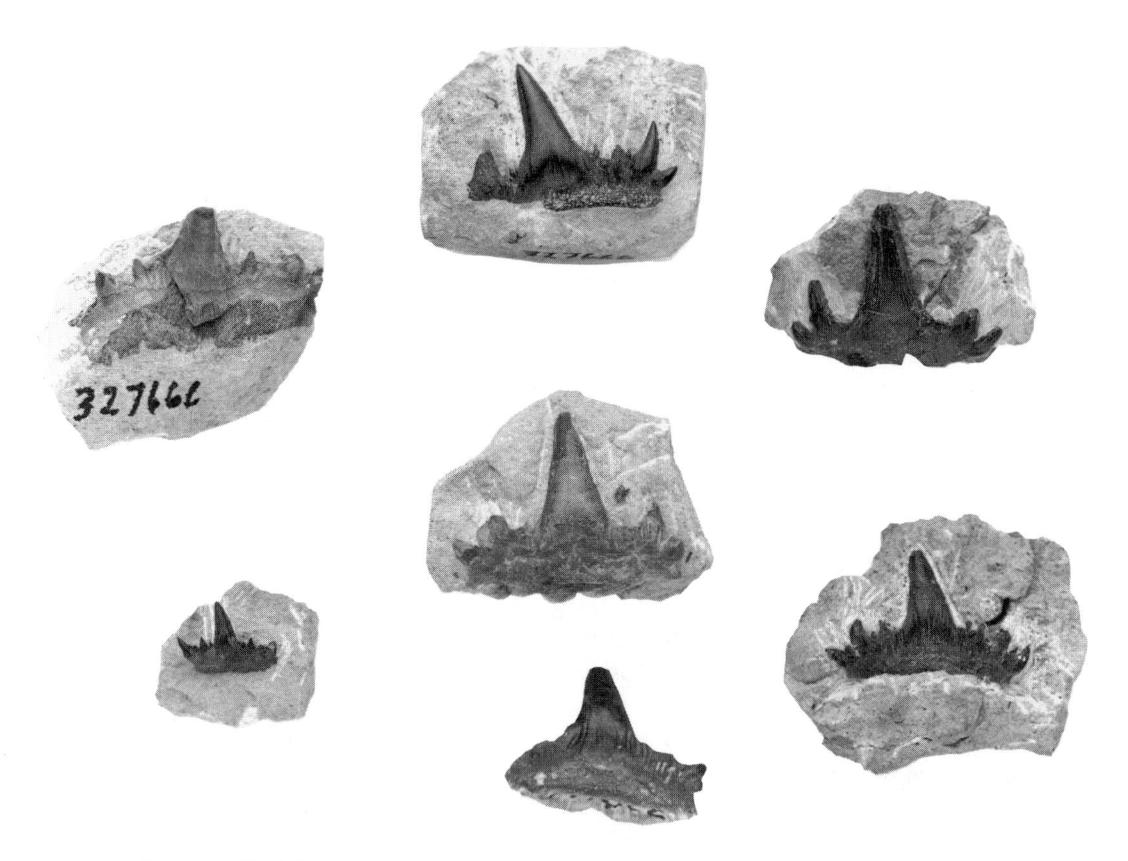

Figure 18. Cladodont type teeth from the Mississippian formation (about 350 million years ago) of Burlington, South Hill, Iowa. (Chip Clark)

known to have been in error (plate 16). A famous, or perhaps, infamous, picture of several scientists and museum personnel standing inside the 2.7 m wide reconstructed open jaws of *C. megalodon* at the American Museum of Natural History in New York City is often cited as evidence that it did indeed grow to such immense lengths. But we now know that the jaws were made at least one-third too large and that all the teeth in the jaws were not the same large size as they are in the reconstruction; the teeth on the sides of the jaws should have been smaller than those in the center. Recent and more reliable studies indicate that the largest fossil *Carcharodon* grew to an estimated length of about 13 m (about 43 ft), which, indeed, is still very large.

Further readings

Cappetta, H., 1987; Compagno, L. J. V., 1973, 1977; Maisey, J. F., 1984a, 1984b, 1985; Moy-Thomas, J. A., 1971; Randall, J. E., 1973; Schaeffer, B., 1967; Schaeffer, B., and M. Williams, 1977.

WHERE ARE FOSSIL TEETH FOUND?

Fossil sharks' teeth (figures 18, 19) are found on the ocean bottom and on land once covered by ancient seas. Sea-bottom deposits that are millions of years old often yield teeth, and, less often, the fin spines of sharks. Fossil skeletons of the earliest sharks are rare, but hundreds have been found in the Bear Gulch formation

Figure 19. Examples of fossil shark teeth from the upper Miocene to recent periods (less than 10 million years ago); related species are still extant: upper left, sand tiger sharks, Odontaspis; *upper right, tiger sharks,* Galeocerdo; *center, snaggletooth sharks,* Hemipristis; *lower left, mako sharks,* Isurus; *lower right, sixgill sharks,* Hexanchus. *Actual size of teeth is 38 percent larger than pictured here. (Chip Clark)*

of Montana, an area once covered by water and dating back 350 million years.

Because sharks' teeth are durable and numerous, they are usually the only evidence of the fish that survives over a long period of time. In the United States, the Cretaceous deposits of the Rocky Mountains and Great Plains, and the Tertiary deposits along the coasts of California and the Atlantic and Gulf states, are areas where shark teeth often can be found. Rich collecting sites are the Calvert Cliffs, located on the western shore of the Chesapeake Bay, beaches of North and South Carolina, and along the coasts of Florida, particularly in the Venice area. Sharktooth Hill in Kern County, southern California, is another abundant fossil site. About 30 species of sharks and rays, dating back 12 to 15 million years, have been identified from Kern County. Some of these teeth are similar to those found along the East Coast. It is not unusual to find the teeth of sand tigers, makos, and, occasionally, cowsharks that have eroded out of the cliffs along the shores of Chesapeake Bay. Because they are avidly sought by collectors, the teeth of *Carcharodon megalodon* (plates 1 and 16), the formidable ancestor of the modern white shark, once common in the Chesapeake Bay area, are increasingly difficult to find.

Fossil sharks' teeth are not unique to the United States. They occur abundantly in all continental areas having formations of the same age as those teeth-bearing formations of the United States. These formations may be well inland from the coast, as they are in Europe, where the teeth occur on land formerly covered by the ancient Tethys Sea, which stretched from the Atlantic through the Mediterranean to the Indian Ocean and beyond.

Colors of the fossil teeth may range from pastel shades to dark gray or ebony. Pale teeth are not necessarily younger fossils than dark teeth, but rather reflect the color of the sediment in which they were buried. Normally, the age of a fossil is based on the age of the deposit in which it is found. Most fossil sharks' teeth are usually less than 20 million years old. They are identified by their shape and size and the location of the site where they were collected. Those found near the Calvert Cliffs are about 12 to 15 million years old.

Further readings
Cartmell, B. C., 1978; Macdonald, J. R., 1983; Purdy, R. W., 1984; Thomas, M. C., 1968; Thompson, I., 1982.

HOW MANY SPECIES OF SHARKS EXIST?

There is no exact answer or complete list of living sharks, although estimates now range over 368 species (the number currently recognized). New species of sharks are described every year. Undoubtedly many others are swimming about in uninvestigated reefs, continental slopes, and the vast and scarcely sampled ocean depths. The sea yields its secrets reluctantly.

One exciting discovery occurred relatively recently, on November 15, 1976. A naval research vessel operating in deep water (4,600 m), northeast of Oahu, Hawaii, deployed two parachutes as sea anchors at a depth of about 165 m. When the chutes were hauled to the surface, one of them had accidentally entangled a large male shark, 4.46 m (almost 15 ft), which subsequently was identified as comprising a new family, genus, and species. Called "megamouth" because of its unusually large mouth, which is lined with a silvery tissue (plate 8: a and b), this shark is the sole member of the family Megachasmidae, whose name is derived from the shark's generic name, *Megachasma* (Greek for large open mouth). Strangely enough, another large, mature male megamouth was caught by commercial fishermen off Catalina Island on November 29, 1984. It was snagged in a gill net set between 7.5 and 37 m in water about 825 m deep. Similar in size to the first specimen, it was estimated to weigh about 700 kg (more than 1,500 pounds). Only these two specimens of megamouth are known.

Not only may the number of shark species increase with the discovery of new forms, the number may decrease when new studies indicate that certain species are really only variants or different growth stages of the same species. In some sharks the juveniles may differ so much from the adults in body proportions and color pattern that they appear to belong to a different species.

Because of the difficulty in capturing certain sharks, there may be only a few specimens available for study. With few data it is small wonder that the identifications of many species of sharks, and consequently the number of species, are still in flux. Recent studies of the requiem shark genus *Carcharhinus* are a case in point. Of the 102 described species referable to *Carcharhinus*, only 28 were recognized as valid (the others were synonyms of the 28); 2 new species were described, and 6 species that had been placed in *Carcharhinus* were transferred to other genera. The assignment of individuals to genera, families, and orders has fluctuated in the past and continues to be the subject of controversy and study among shark specialists.

Further readings

Compagno, L. J. V., 1977, 1981, 1984; Garrick, J. A. F., 1982, 1985; Lavenberg, R. J., and J. A. Seigel, 1985; Springer, S., and G. H. Burgess, 1985; Taylor, L. R., et al., 1983.

HOW ARE SHARKS CLASSIFIED?

Classification tries to bring order out of chaos. Based on relative degrees of similarity in external and internal structure, classification provides a way to catalog the millions of living creatures that inhabit this planet, and provides the framework for comparing and contrasting various groups and for tracing their evolutionary history.

The animal kingdom is divided into phyla that are subdivided (in decreasing order of magnitude) into classes, orders, families, genera, and species. Scientific names are assigned to all animals according to the binomial system of nomenclature established by Carl Linne (or Linnaeus) in 1758. This system is accepted worldwide and consists of two italicized words. The first word is the generic name. It is capitalized and used to group closely related species. The second word is the species name. It is not capitalized for animals and refers to a particular kind of organism included in a genus.

It is generally acknowledged that early in the earth's history the cartilaginous fishes, constituting Class Chondrichthyes, separated into two major subclasses, the Elasmobranchii (sharks and rays) and the Holocephali (chimaeras). Some scientists have subdivided the Elasmobranchii into two orders: Selachii, the modern sharks, and Batoidea, the skates and rays. Others have ranked the Selachii as a superorder composed of four orders, and yet others include the skates, rays, and sharks in four suborders under the Order Euselachi. In this publication we use primarily the classification proposed by Dr. Leonard V. Compagno in 1984, but have modified it to include even more recent studies. We recognize three superorders of living sharks: 1) Squalomorphii, consisting of three orders, six families, and about 24 percent of all living shark species; 2) Galeomorphii, with four orders, 21 families, and about 73 percent of the living shark species; and 3) Squatinomorphii, consisting of one order, one family, and about 3 percent of the shark species. See appendix I for classification.

HOW DO SHARKS SWIM?

Efficiently. Most species are beautifully adapted for locomotion, their streamlined bodies and rigid fins so perfectly designed that they are able to propel themselves through the water with great speed and agility. Sharks have often been compared in structure and hydrodynamics to the shape and aerodynamics of the fighter airplanes of World War II. Anyone who sees a shark swimming must be impressed by its seemingly effortless movements.

In sharks, the tail, or caudal fin, plays a major role in providing the power for swimming. All sharks have a heterocercal caudal fin. In this type of fin the vertebral column extends well out into the upper lobe of the fin, which in most sharks is longer and larger than the lower lobe. The two lobes may also vary in shape, ranging from symmetrical to, more commonly, asymmetrical. Most fast-swimming pelagic (open ocean) species, like the mako, have lobes that are almost equal in size. The length of the tail varies from less than 20 percent of the body length (angel sharks) (plate 12c) to longer than the body length (thresher sharks) (plate 9: b and c). In most sharks, particularly the pelagic species, the axis of the upper caudal-fin lobe is raised (as much as 45 degrees) relative to the horizontal axis of the body, but in sluggish bottom-dwelling species the axis of the fin is in line with that of the body.

Whatever the design, and whether it might be related to the kind of swimming needed for a particular habitat or type of feeding, the tail not only is the major source of power for shark propulsion but also provides for sudden changes in direction. The sweep from side to side, or sculling motion, of the fin produces a thrust forward and upward above the center of gravity.* The upward force at the tail tends to rotate the body about its center of gravity, pushing the head down. This downward motion is counterbalanced by the lift of the pectoral fins and the torpedo-shaped body (and the flattened lower surface of the head, according to some investigators). The combined lift of the pectoral and tail fins and the body keep the shark from sinking of its own weight, while the thrust component of the tail movement propels the shark forward.

The typical fast-swimming pelagic sharks have a rigid first dorsal fin, which acts as a stabilizer, placed well forward on the body. Some sharks also have a smaller second dorsal fin, which also helps to control rolling. Unlike the fins of the bony fishes, which are supported by movable elements and can be folded and waved at will, the fins of sharks are erect and scarcely under conscious control. Except for the flexible, thin, trailing edges of the fins and the sculling and rotating motion applied to the tail, the fins of sharks are not free moving. The pectoral fins, set low on the body, provide a planing surface and lift during swimming. They are used also for steering, braking forward motion, and for help in turning. To turn laterally (yaw), the shark only needs to move its tail asymmetrically (i.e., expend more energy on one side than the other). To climb or dive, it can generate more or less vertical thrust by varying the rotation of its tail. The lateral undulations of the body, the tail movement, and the flexible margins of the pectoral fins all coordinate to achieve remarkable maneuverability; however, sharks, unlike bony fishes, cannot swim backwards.

The shark's skin, particularly the covering of aligned tiny, grooved denticles, increases its swimming proficiency immensely by overcoming drag imparted by the water. Scientists working to decrease drag on the surface of airplanes and boat hulls are experimenting with covering the surfaces of these vessels with a shark-like "skin" having grooves similar in proportions of height and spacing to that of sharks.

Sharks have no air bladder to help adjust their buoyancy as do many of the bony fishes. Although the absence of an air bladder is an advantage for a fish that moves rapidly from one depth to another, it could make it harder to stay afloat. Sharks are denser than water and must develop a certain amount of dynamic lift to avoid sinking. Most sharks solve the problem by constantly swimming, using the tail thrust and planing action of the pectoral fins to maintain position. The shark's liver also helps counteract the natural tendency to sink. This organ occupies as much as 90 percent of the body cavity and 25 percent of the weight in some sharks. It produces lighter-than-water oils that decrease the density of the body and allow

---

*Some researchers suggest (mistakenly, according to our independent sources) that the forward thrust passes *through* the shark's center of gravity (which would cause the shark's snout to pitch upward without any restoring force).

Figure 20. The sand tiger shark, Eugomphodus taurus, *Family Odontaspididae, is about 2 m in total length. This coastal inhabitant lives in all oceans except the central and eastern Pacific and central Indian oceans. (Gregg Skomal, National Marine Fisheries Service)*

species like the basking shark or whale shark to lie almost motionless at the surface. Another way to remain buoyant has been observed in captive sand tiger sharks (figure 20). They come to the surface and gulp air, which distends the stomach and allows them to maintain a motionless position in the water.

Further readings
Blake, R. W., 1983; Thomson, K. S., and D. E. Simanek, 1977; Moss, S. A., 1984; Webb, P. W., 1984.

MUST SHARKS SWIM TO BREATHE?

A common but erroneous idea is that sharks can breathe only when swimming. This is true of many, if not most sharks, but there are some that do not have to swim to breathe. During breathing, a current of water passes through the mouth to the pharynx and over the gills, then passes out the gill slits. The exchange between the oxygen dissolved in the water and the carbon dioxide dissolved in the blood takes place through the surface of the gill filaments. Some sharks do appear to respire by swimming with their mouths open, but others, like the nurse, lemon, and bull sharks, breathe actively, pumping water over their gills by opening and closing their mouths even when not swimming. In many species, usually bottom dwellers, the spiracle, an opening behind the eye, can also pass oxygenated water to the gills.

DO SHARKS SLEEP?

Nobody knows if fishes sleep in the same way that humans do. Human sleep involves changes in eye movements, rates of heartbeat and respiration, and in the electrical pattern of the brain waves. The only way one might conclude that sharks sleep would be to monitor these changes and observe if there were consistently different patterns associated with resting and active sharks. Some bony fishes do remain motionless at times and are apparently resting. Many of these actively feed during the day and lie quietly on the sea bottom at night or seek shelter in caves.

Many, if not most, species of sharks swim constantly, never resting. But some sharks do appear to sleep, or at least enter a torpid state. *Carcharhinus perezi* (plate 4a), the reef shark of the Caribbean, is often encountered resting on the bottom by divers. Ichthyologist Dr. Eugenie Clark, in her visit to underwater caves off Isla Mujeres near the Yucatan Peninsula, observed large carcharhinid sharks, *Carcharhinus perezi*, in a torpid state. These sharks are usually seen actively swimming in the open ocean, but in the caves they were on the bottom, 20 m down, motionless, and apparently so tranquil that one could lift and handle them. Freshwater streams have infiltrated many of the caves, decreasing the salinity of the water and increasing the acidity and carbon dioxide. A relationship may exist between the curious resting behavior of these sharks and the chemical composition of the cave water, or the sharks may be using the infiltrated freshwater to rid themselves of ectoparasites.

Similar behavior has been observed in other sharks, such as the whitetip reef shark, *Triaenodon obesus* (plate 5a), which is commonly seen resting motionless, not only in caves (in Japan) and in crevices in coral reefs, but also on the bottom of aquarium tanks. The whitetip reef sharks ordinarily gather quietly in caves during the day, becoming much more active at night or during periods of slack tide when the currents are less strong. They have been known to interrupt their rest when spearfishing is occurring nearby and they can find a quick meal of a few dead fish.

Further readings

Clark, E., 1969, 1975; Hoar, W. S., and D. J. Randall, eds., 1970, Vol. IV; Moss, S. A., 1984; Russo, R., 1984.

ARE ALL SHARKS GRAY-BLUE?

No, sharks can be colorful. They may have a variety of markings and color both in adult and juvenile forms, but in no way do they match the rainbow hues and extraordinary patterns of the bony fishes. Most sharks are drab colored, and none are silvery or golden, as are many bony fishes.

Countershading, in which the back (dorsal surface) is darker than the underbelly, is common in most fishes, including many sharks. This coloring makes the shark difficult to see from both above and below. Upper-surface colors range from deep indigo blue (blue shark) to shades of olive green and brownish gray (smooth hammerhead) to metallic black hues (thresher shark). Some deep-sea sharks are entirely black, and some, like the lantern dogfish, have luminous organs that glow in the darkness of the depths. The juveniles of some sharks are more distinctively marked than the adults, and have stripes and spots that tend to disappear with age. For example, the tiger shark adult is bluish- or greenish-gray with a light gray to yellow or white belly. The juveniles appear mottled, with dark spots on a paler background (plate 5c). In larger individuals, the spots fuse to form dark bars or a "striped" pattern. In the largest individuals the bars tend to fade, with traces, at most, remaining in the tail region. Dark spots are also present on juveniles of the nurse shark, but absent in adults. The oceanic whitetip shark undergoes a change of color pattern that belies the shark's name: the bright white fin tips of the adult are jet black in the juvenile—a fact unknown at the time the common name was conferred. On the other hand, adult and juvenile whale sharks have the same color pattern, a greenish and purplish to reddish brown, sprinkled with small white or yellow spots and a paler lower surface.

The smalleye hammerhead (*Sphyrna tudes*) is bright orange as a juvenile and yellow as an adult. These colors are probably derived from carotenoid pigments in their preferred foods: shrimp for the young and catfish eggs and fish for the adult.

The most colorful sharks are the cat sharks, almost entirely restricted to the deep slopes of continental coasts, and the collared carpet sharks, wobbegongs, and zebra bullhead sharks (plate 2f), all but a few of which live only in the shallow waters of the Indo-Pacific. Various patterns of spots, stripes, and blotches adorn the bottom-dwelling carpet sharks, providing camouflage for them among the rocks and weeds along the coast. The adult zebra shark, *Stegostoma fasciatum* (plate 2b), for example, is pale yellow with dark brown spots (the juvenile is dark brown with vertical yellow stripes and spots). The collared carpetshark, *Parascyllium collare* (plate 2a), of southeastern Australian coastal waters, sports bold, irregular, dark brown and white wavy bars on its head and body. In Stewart Springer's monograph, *A Revision of the Catsharks, Family Scyliorhinidae,* male cat sharks, *Holohalaelurus punctatus,* hauled up in a trawl off the coast of Tanzania, are described as "one of the most beautiful of all sharks." The back of this shark is orange-brown covered by dark brown spots; the belly is cream colored. Each dorsal fin bears a short, brown bar with a yellow border, and the outer halves of the pectoral and pelvic fins are sprinkled with brown spots.

Further readings
Budker, P., 1971; Springer, S., 1979.

HOW FAST CAN SHARKS SWIM?

Measuring the swimming speed of fishes in the wild is difficult, if not impossible. The only accurate way to measure swimming speed is to clock an animal swimming directly against a current of known speed. Such currents can be established only under laboratory conditions, and many sharks are too large to be placed in laboratory tanks. Those that are not are usually timed only at cruising speeds as it is difficult to induce sharks to swim at high speeds. In captivity, sharks have been

Figure 21. The shortfin mako, Isurus oxyrinchus, *frequently makes spectacular leaps when caught on hook and line. (Charlie Donilon)*

clocked at cruising speeds of 2.0 to 4.8 kmh (1.2 to 3 mph), but it is expected that in the wild, when following prey or attacking, those speeds are increased tenfold. The larger the shark, generally, the faster it can swim, particularly when making an attack. Mako sharks, which are among the most streamlined, have been observed jumping as much as 6 m out of the water (figure 21). Calculations show that to do this the mako would require a starting velocity of about 35.2 kmh (22 mph).

In one set of experiments, which have been questioned, a blue shark 0.6 m long and weighing 0.6 kg was able to maintain its speed against a current of 7.92 m per second (17.7 mph), and in short bursts, similar to an attack situation, it attained a speed of 19.21 m per second (43 mph). Lemon sharks are known to achieve brief (two seconds) bursts of speed of up to 30 kmh. They have been chased for up to 4 to 4.5 minutes at speeds of 17 to 18 kmh, after which the sharks "run out of gas" (note that these speeds are not corrected for the effects of currents). Normally, lemon sharks cruise at about 2.5 kmh.

Further readings
Budker, P., 1971; Lineaweaver, T. H., III, and R. H. Backus, 1970; Weihs, D., et al., 1981.

DO SHARKS MAKE SOUNDS?

Not normally. Sharks have no sound-producing organs and seem incapable of making sounds. Divers have reported, however, that the whale shark, largest of all fishes, makes a constant croaking sound. The swell shark is able to inflate its belly by gulping water, a technique that may make it difficult for a predator to remove the shark from a hiding place with a small opening or in which the shark is wedged. Out of water, the swell shark is able to inflate its belly by gulping air, and according to reports of New Zealand fishermen, expel the air with such force as to make a dog-like barking sound. Another example of a similar type of sound production is exhibited by the spitting shark, *Nebrius ferrugineus,* of the Indo-Pacific, named for its habit of expelling water from its mouth in spurts when captured. The spitting shark reportedly accompanies each "spit" with a coughing grunt. Presumably, the horn sharks, which have plates of heavy teeth in their jaws, make sounds as a byproduct of their feeding on hard-shelled mollusks, which they crush before swallowing.

Further readings
Grant, E. M., 1982; Springer, S., 1979.

HOW LARGE DO SHARKS GROW?

Sharks never stop growing, so finding their size limit requires knowing their age limit—and neither is easy to do. (See the next question, "How long do sharks live?") Reasonable estimates of sharks' sizes have been made, however, and no other aspect of sharks so vividly illustrates the tremendous variation among species as does the range in recorded sizes. At one extreme is the newly described (1985), deep-dwelling (almost 300 m, about 1,000 ft) dwarf dogshark, *Etmopterus perryi* (plate 3a), from the Caribbean sea off Colombia. It is not the smallest fish, but is perhaps the smallest of all the sharks, males attaining a length of only 16 to 17 cm (6.4 to 6.9 in), and females only 19 to 20 cm, at sexual maturity. The largest embryos known for this shark are only 6 cm (2.4 in) long.

Several other deep-dwelling species of sharks are almost as small at maturity as *Etmopterus perryi,* but there are relatively few small, shallow-dwelling species. Among the smallest of the shallow dwellers is the Indo-west Pacific spadenose shark, *Scoliodon laticaudus* (figure 22). Spadenose sharks are about 13 cm (5 in) long when born. Males reach sexual maturity at about 35 cm, and females at a somewhat larger size. The largest reliably reported spadenose sharks are less than

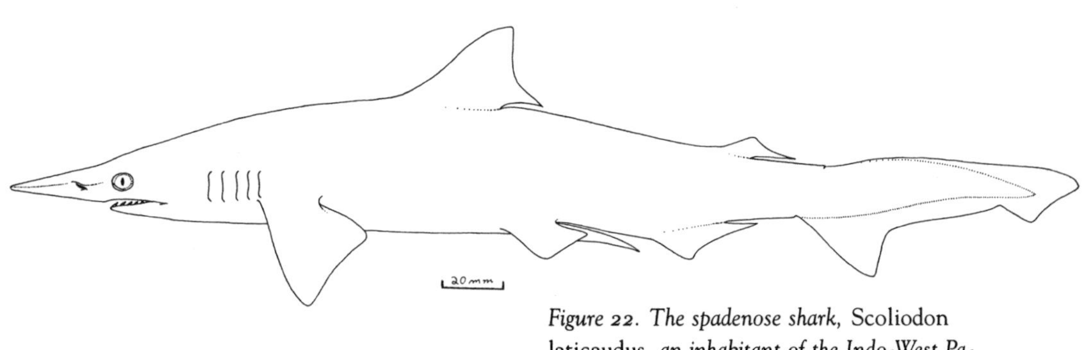

Figure 22. The spadenose shark, Scoliodon laticaudus, *an inhabitant of the Indo-West Pacific, is one of the smallest shallow-dwelling sharks. This species is free-living at a length of about 13 cm, smaller than illustrated here, and attains a length of about 75 cm. (After Springer, 1964)*

75 cm. A species of Japanese cat shark, *Scyliorhinus torazame,* attains a maximum length of only 50 cm.

At the other end of the size range is the nonpredatory, shallow-dwelling, circumtropically occurring whale shark, *Rhincodon typus* (frontispiece), the largest fish known, living or fossil. This species has been reliably measured at over 12 m, which is one and a half times longer than a standard 54-passenger American school bus, but is known to grow much larger, possibly to 18 m (59 ft). Such huge fishes are almost impossible to weigh; suitable scales and equipment often are not available, but it has been estimated that one 11.6 m specimen weighed about 12,088 kg (26,594 lb, or over 13 tons). The world's second-largest fish is also a shark. The basking shark, another nonpredatory, shallow-dwelling species, but living in cooler waters (both it and the whale shark feed on plankton), has been measured at 9.75 m (32 ft), although lengths of 12.2 m to 15.2 m are attributed to it. One specimen, measured at 9.1 m, was estimated to weigh 3,900 kg. The largest reliably recorded specimens of the most feared of all sharks, the white shark, have measured from 5.8 m to 6.4 m and weighed 3,324 kg, although a questionable report of a 9 m specimen from the Azores is given currency in the popular literature. The largest deep-dwelling sharks include the predatory Greenland shark, *Somniosus microcephalus,* which attains at least 6.4 m, and the filter-feeding megamouth shark, at 4.6 m.

In between the two extremes of size are the vast majority of sharks. Although a few species other than those mentioned may attain lengths of from 3 m to 6 m, over half the species are less than 1 m long at their maximum, and about 80 percent do not exceed 2 m in length (see appendix III).

Further readings

Ellis, R., and J. E. McCosker, 1986; Springer, S., and G. H. Burgess, 1985; Springer, V. G., 1964; Tricas, T. C., and J. E. McCosker, 1984.

HOW LONG DO SHARKS LIVE?

Some sharks may live for almost a century, but the average life span is less than 25 years. Most of what is known about the life span of sharks comes from aquarium records of sharks in captivity or from tagging and recovery programs. These data show that the maximum age attained varies according to the species.

The spiny dogfish, Methuselah of the shark world, has been reasonably estimated to live more than 70 years, perhaps even 100 years. The Australian school shark, *Galeorhinus galeus* (see figure 44) also has a long life. This shark is well known because it is fished commercially and has been used as food for almost 60 years in southeastern Australia. Different estimates have been given for the age attained by the school shark. It has been estimated to live at least 53 years, based on an individual that was caught 33 years after it was tagged and believed to be 20 years old at the time of tagging (figures 23 and 24). Another school shark recently caught off Tasmania's east coast had carried a tag for 34 1/2 years. This shark was thought to be 7 to 8 years of age at the time of tagging, so that at capture it would have been over 41 years old. Age estimates for the school shark at the time of tagging were based on the length of the shark and what had been learned about growth from other tagged specimens. Sharks are measured, tagged, released, recaptured, and remeasured to find out their rate of growth—so many cm in so many years (plate 13: a, b, c, and d).

Another species probably long-lived is the whale shark. Age estimates for this huge beast are as high as 60 years, but no one really knows how old whale sharks get to be. The life spans of most of the large, dangerous species are also unknown. It has been impractical, costly, and dangerous to keep these sharks in captivity, or to tag them. Most shark species cannot adapt to captivity, refuse to eat, and die soon after capture. All attempts to keep a white shark in captivity (there have been more than a dozen) have failed. One white shark, placed in a tank at Marineland, Florida, died 35 hours after capture. Another, maintained at San Francisco's Steinhart Aquarium for three and a half days, had to be released after she became disoriented and persisted in colliding with a metallic part of the tank.

No one method for calculating age and growth is reliable for all species of sharks. In addition to tagging and recapture records, another method commonly used is to count the growth rings (or annuli) on skeletal hard parts, such as the vertebral centra (plate 13: e and f). In species such as the scalloped hammerhead, it seems that rings form once a year. In others, like the dusky shark, more than one ring may form annually, but rings in angel sharks seem not to be related to monthly or yearly time periods. Age determinations for only a few species, such as the shortfin mako, Atlantic sharpnose, spiny dogfish, bull shark, and blue shark, have been made using the annual ring method. Annual rings are more apt to form in sharks that live in seas having wide seasonal changes in temperature. Tropical seas have more constant temperatures, and species in these areas may not form long-term periodic marks on skeletal structures.

M·91364

TO ASSIST BIOLOGICAL RESEARCH, KINDLY SEND THIS LETTER TO ADDRESS BELOW, WITH SPECIES, LENGTH AND WEIGHT OF FISH AND DATE, LOCALITY AND METHOD OF CAPTURE. REWARD. DATA ON RELEASE OF FISH WILL BE SENT. PLEASE RECORD NUMBER.

PARA AYUDAR A INVESTIGACIONES BIOLÓGICAS, HAGA EL FAVOR DE MANDAR ESTA CARTA A LA DIRECCIÓN DE ABAJO, CON INFORMACION DE LA ESPECIE, LONGITUD Y PESO DEL PEZ, Y FECHA, LOCALIDAD Y METODO DE CAPTURA. PREMIO. APUNTE USTED EL NÚMERO.

NATIONAL MARINE FISHERIES SERIVCE LABORATORY, NARRAGANSETT, RHODE ISLAND U.S.A.

M·91364

Figure 23. (Above, left) Dart tag (actual size) with separated example of instructions to finder contained in sealed capsule at the free end of the tag. This is the tag most commonly used on sharks. (H. Wes Pratt, Jr., National Marine Fisheries Service)

Figure 24. (Below) Row tag (magnified 1-1/8 ×), is obsolete. Sharks bearing this type of tag have been caught as many as 15 years after being tagged. Although the tag number may become illegible, useful information can be obtained. Anyone who catches a tagged shark should immediately notify the National Marine Fisheries Laboratory, Narragansett, Rhode Island 02882; phone (401) 782-3200. (H. Wes Pratt, Jr., National Marine Fisheries Service)

One way of checking if vertebral growth rings are formed annually is to inject a captive living shark with the antibiotic tetracycline. Tetracycline leaves deposits in growing tissues containing calcium and can form a more-or-less permanent marker that fluoresces under ultraviolet light. A shark injected with tetracycline, released, and recaptured can have its vertebrae checked to see how many growth rings formed outside the tetracycline marker. If the shark was free for three years and three rings formed, one could safely conclude that a ring had been formed each year.

Having established the annual nature of ring formation experimentally in a few individuals of a particular species of shark, one could examine a large number of individuals of the same species that were not marked with tetracycline and plot a graph of shark lengths versus number of rings in the vertebrae. If length increases in a more-or-less regular way, one can make a reasonable guess of the age of an individual shark of that species by knowing only its length. One other piece of required information is the animal's age or size at formation of the first growth ring. This can be determined by examining the vertebrae of near-term embryos and the smallest known free-living individuals of the species in question. Vertebrae from such individuals normally do not have growth rings, thus, one could assume that any individual of that species with a single growth ring was less than 2 years old.

Another way of estimating age is by the use of length-frequencies. If birthing of a species of shark occurs only during a brief period of the year, for instance, in the spring, all the individuals born in the spring of a particular year will be of a limited size range, and individuals belonging to this group (called a "year class") will grow at about the same rate. If many sharks of widely different sizes but the same species are caught during a limited period of time, say June of a particular year, a graph of the sizes of these sharks should show a series of peaks. Each peak would represent the modal (most frequent) size for each year class. For instance, if five peaks occurred, these would represent individuals from less than one to four years of age. Since it may not be possible to obtain a sufficient sample of sharks in a single month, one might lump all the data for the month of June, for example, for several years. Even under ideal circumstances, however, identifying more than the first five to ten peaks is difficult, because as sharks (and other fishes) grow older, growth rate tends to slow down and the peaks fuse or become indistinct. Aging by the length-frequency method generally has been unsuccessful because of the difficulty of obtaining adequate numbers of any shark species throughout its size range during each month or season of the year.

In one study on the sandbar shark, *Carcharhinus plumbeus* (figure 25), age and growth were estimated using most of the methods described above. These methods were used to cross check each other and involved some complex mathematical analyses. The results indicated that sandbar sharks may live for over 30 years. (In 1986, three tagged sandbar sharks were recaptured after having been at liberty for 20 to 21.3 years.) The growth rate is similar for both sexes, although the adult female attains a larger size than the male. Up to the age of 3 years, growth is relatively fast. A slowing occurs between ages 4 through 9, after which the rate

Figure 25. The sandbar shark, Carcharhinus plumbeus, *found circumglobally along warm coasts, but in the central Pacific, known only from the Hawaiian Islands, where it is the most common species around the principal islands. The individual illustrated (1.79 m) is from the Red Sea, where the species has a higher first-dorsal fin than is found in individuals from other areas. (John E. Randall)*

increases for a few years, then appears to decrease and remains constant until death. No one knows why the growth rate changes, although factors such as movement to a different habitat or a shift in food habits may be responsible.

The most extensive field studies of the lemon shark, *Negaprion brevirostris,* are being conducted in the Florida Keys and off the Bimini Lagoon in the Bahamas by a team of researchers led by Dr. Samuel Gruber. Lemon sharks have been captured, marked and tagged, injected with tetracycline, and fitted with radio transmitters that emit trackable high-frequency sounds. This research has led to changes in the estimates of the growth rate of the lemon shark. It was once thought they grew rapidly, reaching maturity in one or two years. Now we know that even though they are long lived, reaching 50 or more years of age, they grow very slowly, increasing in length and weight by only 0.4 m (16 in) and 3.6 kg (8 lb) after four years in the nursery areas. They also mature late, not mating until they are at least 12 years old, and probably not before they are 15 to 20 years old. At that age they may weigh about 114 kg and are about 2.4 m long.

Further readings
Aasen, O., 1963; Cailliet, G. M., et al., 1983; Casey, J. G., et al., 1983, 1986; Castro, J., 1987; Gruber, S. H., 1985, 1988; Gruber, S. H., and R. G. Stout, 1983; Morrissey, J. F., 1987; Nelson, D. R., 1978; Olsen, A. M., 1984; Parsons, G. R., 1985; Pratt, H. L., Jr., and J. G. Casey, 1983; Schwartz, F. J., 1983; Taniuchi, T., et al., 1983.

WHERE DO SHARKS LIVE?

All the oceans of the world, except possibly the coldest waters of the Antarctic, are home to one or more species of sharks. They inhabit warm tropical and subtropical waters in greatest numbers, but are also found in the icy waters of the Arctic. They live in shallow coastal areas and the open ocean. They may be confined to near-surface waters or to the abyssal depths, where the only light is that cast from their own luminous organs or those of other creatures. A few species inhabit the brackish water of estuaries and the freshwater of tropical lakes and rivers, and some of these migrate from saltwater many kilometers upstream to freshwater. Sharks do not occur in European freshwaters or in rivers of northern North America and northern Asia. They are not known in the Great Lakes of North America or in any of the truly landlocked marine or freshwater lakes of the world, including the slightly brackish Caspian Sea of Russia. Only three small, harmless species occur in the Black Sea: the cat shark, *Scyliorhinus canicula,* and dogfishes, *Squalus acanthias* and *Squalus blainvillei.*

Ichthyologists (scientists who study fishes) have been interested in the sharks that occur in freshwater. Formerly, the sharks found in Lake Nicaragua, Lake Izabal in Guatemala, the Ogowe River in western Africa, the Zambezi River in eastern Africa, the Tigris River near Baghdad, and the Amazon River of South America, to name but a few, were each believed to be a different species that was confined to freshwater. Recent studies, however, demonstrate that in all these rivers and lakes there is but one species: the bull shark, *Carcharhinus leucas* (figure 26), which is also commonly found in the warm, shallow waters of the oceans. In Lake Nicaragua, which drains into the Caribbean Sea by the broad Rio San Juan, the bull shark apparently moves freely between saltwater and freshwater. Even the rapids

Figure 26. The bull shark, Carcharhinus leucas, *considered by some ichthyologists to be the most dangerous of all sharks, is found in all warm continental coastal waters, and swims far up freshwater rivers. It has been recorded from the entire central Pacific (Society Islands) only once. This small female, 0.51 m, was taken off* Cochin, India. *(John E. Randall)*

along the river seem to present no barrier to its passage. The bull shark has been implicated frequently in shark attacks on bathers, especially in warm freshwaters such as Lake Nicaragua and the Zambezi River, which have access to the sea.

There are still questions about the sharks that occur in the Ganges River of India. Although no one has reported seeing a bull shark in that river, a newborn bull shark was collected from the Hooghly River, westernmost of the many channels that lead to the Ganges. One might expect, therefore, that the bull shark does occur in the Ganges. Dr. J. A. F. Garrick, author of a recent monograph, *Sharks of the Genus Carcharhinus,* which includes the bull shark, believes that the shark usually identified as *Carcharhinus gangeticus,* despite a superficial resemblance to the bull shark, actually belongs to a different genus. *Glyphis gangeticus* is the scientific name used for the Ganges River shark in L. V. J. Compagno's recent classification, and it is possible that both it and the bull shark coexist (or existed) in the Ganges sytem.

HOW DO SHARKS REPRODUCE?

Sharks reproduce in a variety of ways, but all are accomplished by internal fertilization. The process of internal fertilization involves complex internal and external reproductive organs (figure 27).

The external reproductive organs of male sharks are the pelvic-fin claspers described earlier. The pelvic fins of the female have no special modifications.

The internal organs of the male (figure 27b) include an elongate testis on each side of the body cavity. Each testis, which lies close to an elongate kidney, produces sperm. The sperm from the testis passes through fine ducts that connect to special kidney tubules through which the sperm is led into a larger tube, the ductus deferens. Each ductus deferens not only collects the sperm but also seminal fluid produced mainly by a gland that is formed at the front end of each kidney. The sperm and seminal fluid together are called semen. The semen passes through the enlarged posterior end of each ductus deferens, where it is received into a storage reservoir called a seminal vesicle. The two seminal vesicles meet posteriorly and form a common compartment, the urogenital sinus, that leads into a tiny tube, the urogenital papilla. When mating begins, the semen passes into the urogenital sinus, then out the urogenital papilla into the cloaca. A cloaca is present in both sexes, and is a collecting compartment that opens externally between the pelvic fins. Internally the cloaca has openings from the digestive, urinary, and reproductive systems.

The female reproductive system consists of paired ovaries and oviducts (figure 27a). In many sharks, if not most, only one ovary, usually the right, produces eggs. In some, only the left, and in others, both ovaries are functional. The nonfunctional ovary usually atrophies or is reduced in size. An oviduct consists of four regions: 1) an ostium or funnel, which receives eggs released from the ovary; 2) a

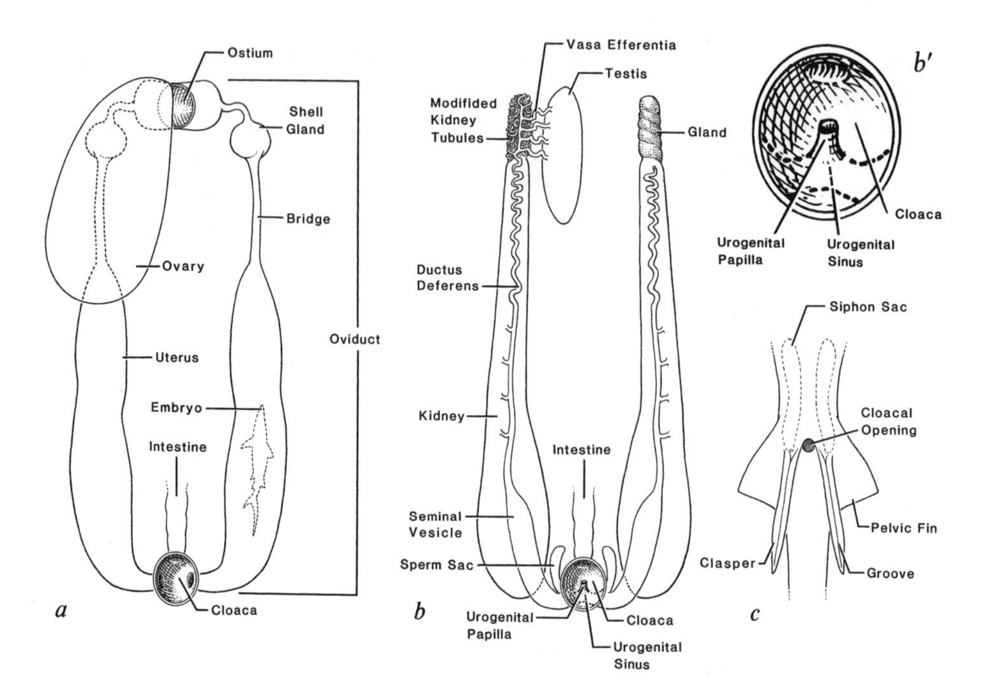

Figure 27. Schematic drawings of shark reproductive anatomy: a) female (one ovary removed); b) male (modified kidney tubules and testis removed from one side); b′) enlarged view of male urogenital region; c) clasper and siphon sac.

nidamental or shell gland, which secretes a protective case over the fertilized egg and stores sperm in some species; 3) a slender bridge or pathway to the posterior part of the oviduct, which is expanded as 4) a uterus that acts either simply as a passageway for the fertilized eggs on their way to being laid, or as an area in which the embryos develop until they are ready to be born. The oviducts, at their posterior end, open either singly or together into the cloaca. The ovary does not connect directly with the oviducts, both of which are present and functional.

During mating, the male presses a clasper against his cloaca, causing the semen to pass into a deep groove on the clasper (plate 15c). One clasper is rotated forward and inserted into the cloaca of the female. The next step in transferring the semen to the female involves the siphon sac, a structure in the abdominal wall, near the front end of each pelvic fin of the male (figure 27c). The siphon sac opens into the groove of its respective clasper. The sac contains a fluid consisting of some glandular secretions from the wall of the sac, but mostly seawater, which is pumped into the sac from the outside by the shark flexing its clasper. Muscular contraction of the siphon sac forces the fluid from the sac into the clasper groove and washes the semen into the cloaca of the female. From there, the sperm enter the female's reproductive tract, where fertilization occurs.

The inserted end of a clasper can expand, exposing hook-like structures that hold the clasper in place during mating.

In the female, the mature eggs leave the ovary and pass directly into the female's body cavity near the ostium. They are swept into the ostium by the movement of cilia, hairlike cells distributed over various parts of the viscera and body cavity. Fertilization occurs in the ostium at the anterior end of the oviduct. After entering the cloaca of the female, the sperm are either stored in the shell gland for later use or go directly through the oviducts to unite with the mature eggs. (Females of the blue shark may mate before their eggs are mature and store the sperm until their eggs are ripe, after which fertilization occurs.)

Courtship and Mating

Courtship behavior in sharks is strenuous and, in some species, often violent. Males nip, bite, and harass the female to gain her cooperation. A holding bite on the female's pectoral fin, such as occurs in the whitetip reef shark, keeps the copulating pair close together. The presence of tooth marks on the gills, head, and back of a female, therefore, suggests she has probably mated. Although the male seldom inflicts mortal wounds, he may himself fall victim to attack by the larger, more muscular female (females of many shark species grow larger than males). For the large species, almost nothing about mating is known, and the only inferences are from evidence such as scars. Where, when, and how often mating occurs in the ocean are mostly unknown.

Most documented reports of reproductive activity are for a few of the smaller species, mainly those held in marine laboratories and aquariums. Just as there are variations in the form of sharks, so there are variations in their mating activities.

In December of 1949, visitors to the Steinhart Aquarium in San Francisco were dismayed to see what appeared to be a fight between two sharks. In fact, they were witnessing the mating of the horn shark, *Heterodontus francisci.* A male horn shark, after chasing a female about the tank, seized her by the pectoral fin with his mouth and held on in an effort to manipulate her body into position to copulate. The female lay partially on her left side while the male placed his tail over her back so that he could insert his right clasper into her cloacal entrance. Once subdued, the female made no attempt to escape, remaining immobile during the 35 minutes of copulation. The same male was seen at other times to chase other females and bite and hold onto them in the gill area.

One of the few sharks actually observed mating in the wild is the cat shark, *Scyliorhinus canicula.* The observed male looped himself around the female and formed a tight coil around her pelvic region while thrusting his left clasper into her cloaca. The couple remained entwined and quiet for about 20 minutes even though harassed by other male sharks that were tugging at the female's tail and assaulting her snout. The male's eyes were closed throughout the activity, and only after completing copulation did he open them, twitch his body, and allow the female to escape, after which he remained twisting and turning on the sea bed for some time.

The only large species of shark that has been seen mating (in captivity) is the lemon shark. The male and female swam side by side with a rhythmic synchronous movement, their heads slightly apart and the posterior portions of their bodies and tails touching closely. This may be the mating position of other large shark species.

The mating of nurse sharks is similar to the parallel swimming seen in lemon sharks, usually with the male slightly behind the female and including, at times, a third individual of unknown importance. From time to time, the male bites the female's pectoral fin, which seems to stimulate her to roll over. In this upside-down position, the female stays still while the male manipulates her body so he can insert a single clasper. Accomplishing this, he rolls over on his back and lies motionless beside her on the ocean floor for about 2 minutes. Then he disengages, and both swim away.

In some species, visual and chemical signals appear to play important roles in precopulatory behavior. Mature male blacktip reef and whitetip reef sharks are known to trail mature females of their respective species closely, within 30 cm (about 1 ft), for periods of tens of seconds. The females, with their tails slightly uplifted, lead the males. Visual cues and an olfactory signal, such as the release of a sex attractant substance (pheromone) by the female, may provide information on her whereabouts and reproductive readiness.

Development and Birth

Most bony fishes produce large numbers of eggs at each spawning, often many thousands, which are deposited in nests or released freely into the water, to be fertilized by the males. The fertilized eggs and resulting young are tiny, defenseless, and vulnerable to predation. Sharks are quite different, producing relatively few eggs during a single breeding season, all of which are retained by the female, at least until they have been fertilized by the male.

All shark fertilization occurs internally, and in most sharks the developing shark embryos are safeguarded inside the female's reproductive tract, where, according to the species, they may develop freely, be attached by a placenta, or be sealed in leathery egg cases. A few shark species deposit the egg cases on the bottom or in vegetation. Depending on the species, a female may bear only a single pup (as young sharks are called) or as many as 100. The pups, which often look like miniature adults, are large, which increases their ability to avoid predators, defend themselves, and find food.

Sharks that lay egg cases with developing embryos are called oviparous (literally, egg birth, or born from an egg). Those that bear young that are nourished by the mother through a placental attachment are called viviparous (born live), and those that bear young that develop internally but are unattached to the mother, including those in egg cases, are called ovoviviparous (born live from an egg). The developing young of ovoviviparous species rely entirely or almost entirely on their own egg yolk for nourishment. The vast majority of shark species are ovoviviparous, giving birth to fully developed, nonplacentated young.

Among the ovoviviparous species, the developing young may be nourished not only by the large egg yolks but also from fluids secreted by the lining of the uterus. The embryos of some sharks, such as those of the sand tiger and mako, also obtain nourishment by eating their own potential "brothers and sisters," both the unfertilized eggs and unborn embryos. This intrauterine cannibalism is known as ovophagy (literally, egg eating). In the sand tiger and mako, apparently only one embryo in each of the paired uteri survives the cannibalism. The voracious nature of the sand tiger embryo requires anyone dissecting the shark to exercise care. Some scientists have been bitten by the unborn embryo, which has sharp teeth and can move about in the uterus. Porbeagle embryos, as well as those of the thresher, and possibly even the basking shark, also engage in uterine cannibalism, devouring many of their siblings (plate 14: b and c).

The spiny dogfish has modified its ovoviviparous mode of reproduction to include two different developmental stages. The early developing dogfish embryo is contained in a thin, amber-colored, membranous egg case, which looks like, and is called, a "candle" (plate 14a). A candle may contain several embryos (figure 28). After 4 to 6 months, the protective case, or shell, ruptures and/or dissolves (the remains of a case have not been found in the uterus), but the embryos continue to develop within the female, receiving nourishment from a large sac of yolk (or yolk sac) suspended from their bellies (figure 29). Spiny dogfish sharks have a long gestation period, 22 to 24 months, perhaps the longest known for any vertebrate. By comparison, the longest gestation period for a mammal, the African elephant, is about 22 months. Adult spiny dogfish have sharp fin spines, which in the embryos have knobs at their tips, presumably so the mother will not be hurt when the newborn are expelled in a series of contractions, head first from the uterus. After birth, the knobs drop off.

Viviparity is the least common, but most specialized mode of reproduction in sharks, occurring in hammerheads, blue sharks, bull sharks, and some of the smooth dogfishes (plate 15: a and b). The embryos of these sharks lack the large

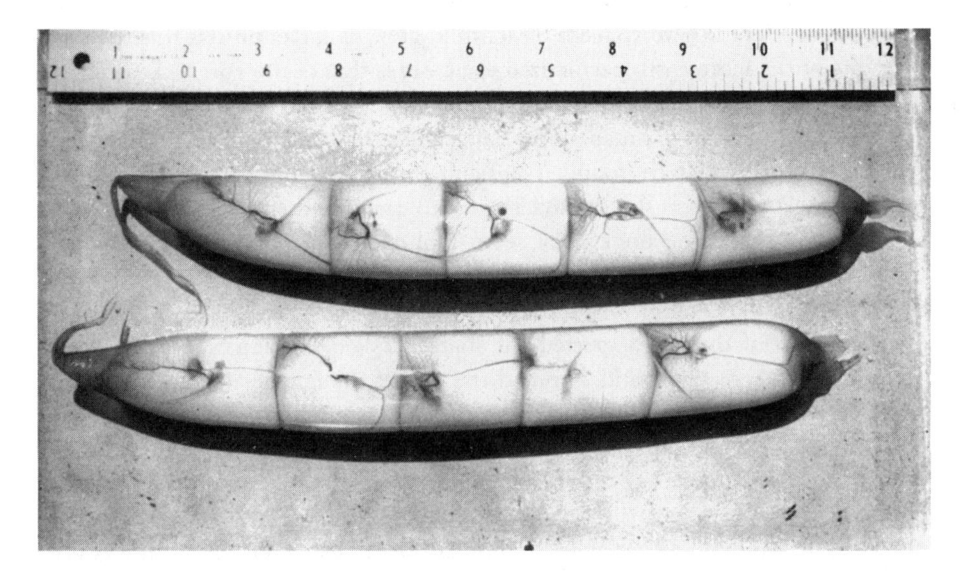

Figure 28. Two spiny dogfish, Squalus acanthias, *candles, each with five embryos. The candle wrapper will rupture, releasing the embryos, which will complete their development in the uterus of the female. (Perry W. Gilbert)*

Figure 29. Spiny dogfish embryo with yolk sac, from which the embryo obtains its nourishment (Ken Lucas, T. F. H. Publications, Inc.)

quantity of yolk that accompanies the embryos of oviparous and ovoviviparous sharks. Instead, the yolk sac is modified into a placenta, which is in close contact with the mother's uterine wall and connected to the embryo by an umbilical cord. Nourishment and oxygen are diffused through the uterine wall to the placenta and then to the embryo. When born, the young may bear umbilical scars, something like the navels of mammals, but they lose them soon after birth.

Oviparous species such as the swell shark, horn shark, many of the cat sharks, and possibly the whale shark, lay their fertilized eggs individually encased in hard, resistant, leathery capsules. These egg cases sometimes wash ashore and since ancient Greek times have been called "mermaid's purses" from their resemblance to coin purses of the time. They vary in size and shape, but are usually either conical or rectangular with long spiral filaments, called tendrils, extending from each of the four corners of the case (figure 30). The tendrils help secure the case to strands of corals, weeds, or rocks, where the cases remain hidden from enemies while the developing embryo feeds off its yolk, growing larger until it fills the case.

One of the more easily recognized egg cases is that of the Port Jackson shark, *Heterodontus portusjacksoni* (figure 31). It is screw-shaped, 13 cm to 17 cm long, and 5 cm to 7 cm wide with broad spiral flanges that help wedge it securely in rock crevices. Even divers have found it difficult to unscrew the case from its mooring. Female Port Jackson sharks have been seen carrying their egg cases, perhaps a form of parental care, since they seem to hide them so well. Development of the embryos takes 9 to 12 months, after which the young hatch and move into nursery areas.

Egg cases of the small-spotted cat shark, *Scyliorhinus canicula,* vary in size according to where the sharks live and the size of the female. In the Mediterranean, egg cases are about 4 cm long and 2 cm wide, whereas those from the eastern Atlantic are 5 cm to 7 cm long and 2 cm to 3 cm wide.

Even though a large egg case belonging to a whale shark was found in the Gulf of Mexico in 1953, no one is sure whether the shark is indeed oviparous. The case measured 30 cm long, 14 cm wide, and 9 cm thick, lacked tendrils, and was very thin. It contained a nearly full-term, 36 cm embryo. Because the smallest free-living whale sharks known are 55 cm and have evidence of umbilical scars, the egg case may have been aborted and the whale shark is probably ovoviviparous.

In the swell shark, development takes as long as a year, the embryo feeding only on its large yolk. When the capsule splits open, the newborn pup is ready to hunt and eat its normal diet of crustaceans and mollusks.

When females of some species are ready to lay eggs or give birth, they may migrate to "nursery" areas where the waters are warm, shallow, and abundant with small fish and other kinds of food. At this time, the females do not feed, which ensures the safety of their pups from cannibalism, a common behavior of sharks. No other parental care is evident, and soon after giving birth the females return to

Figure 30. Egg case of the swell shark, Cephaloscyllium ventriosum, *with embryo inside. Tendrils at ends of the egg case become entangled in marine vegetation, hiding the egg case and preventing it from being carried into deep water or washed ashore, where it might be destroyed. (Marineland of the Pacific)*

Figure 31. Egg case of the horn shark, Heterodontus francisci. *This egg case is very similar to that of the Port Jackson shark. (Steinhart Aquarium)*

the open sea and resume their normal feeding. The pups may stay in the nurseries until they have grown considerably larger and are more mature.

Along the central east coast of Florida, female bull sharks, *Carcharhinus leucas,* normally residing in deep waters offshore, enter the shallow Indian River lagoons between May and July to give birth. Apparently they return to the sea shortly afterward, leaving behind their newborn pups to feed on stingrays and marine catfishes, which seem to be their preferred food. It is not known if the pups leave the lagoons the first winter, but if they do, they return and continue to feed and grow until they approach sexual maturity, which takes perhaps as long as 3 years. Juveniles up to 180 cm (72 in) are found in the lagoon waters throughout the year, and they are also present along the ocean beaches. Because the presence of mature males is rare in the lagoons and rivers of east Florida, mating, believed to occur in June or July, may take place elsewhere.

In North America, sandbar sharks, *Carcharhinus plumbeus* (figure 25), are found in coastal waters from New England to the Gulf of Mexico. When the females are ready to give birth, usually from March to early August, the sexes segregate themselves. The males move to deeper waters, leaving the females in the shallow coastal waters and estuaries. The newborn pups and young stay in the nursery areas until cooler weather, when they move offshore to deeper water.

Further readings
Clark, E., 1963; Dempster, R. P., and E. S. Herald, 1961; Dodd, J. M., 1983; Gilbert, P. W., 1981/82; Gilmore, R. G., et al., 1983; Johnson, R. H., and D. R. Nelson, 1978; Klimley, A. P., 1980; Nammack, M. F., et al., 1985; Pratt, H. L., Jr., 1979; Snelson, F. F., Jr., et al., 1984; Wourms, J. P., 1977.

HOW DO SHARKS DETECT PREY?

Perhaps no other organism has as diverse an array of complex sensory mechanisms to detect prey as do sharks. They can see, smell, hear, feel, touch, taste, detect vibrations and movements (perhaps, even those of the earth), and sense electric fields (figure 32). The image of the shark as primitive or mindless is a widely held misconception. It is not known, however, how the shark's sensory systems work in concert with one another, or the order in which the senses are programmed to seek and devour prey.

There are many stimuli in the ocean, but the most important to a shark seeking food is probably sound. Sharks have an acute sense of hearing and use it to locate prey and, perhaps, other competitors and predators. Sound travels faster and farther in water than it does through air. Even at distances of more than a mile, sharks are attracted to irregular, low-frequency vibrations (particularly 40 Hz or below) such as the struggling or splashing a fish makes when it is wounded. Pure tones and continuous sound do not appear to attract free-ranging sharks. Sharks lack external ears but do have an inner ear on each side of the head that is

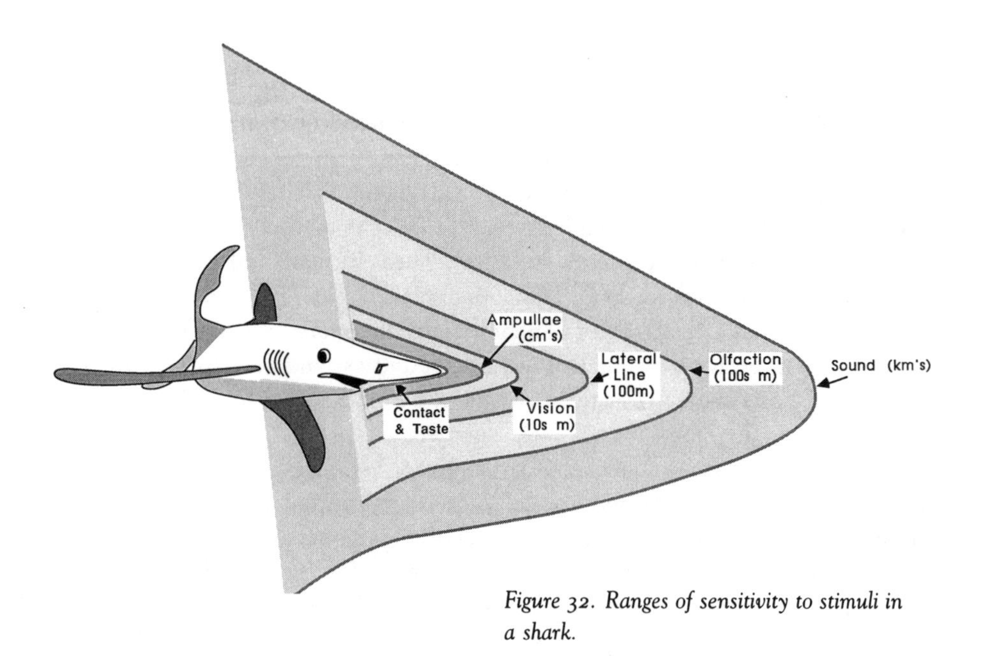

Figure 32. Ranges of sensitivity to stimuli in a shark.

connected by a tiny duct to the outside. Sharks also have a lateral-line system, which extends along the sides of the head and body. The lateral-line system consists of a series of fluid-filled sensory canals containing tiny hairlike receptors that are sensitive to vibrations, pressure changes and waves, and movements of the water. The lateral-line system aids the shark in maintaining its balance as well as detecting sound.

Sharks have a keen sense of smell. Their nostrils perform solely as olfactory organs, playing no part in the breathing process. Sharks scent food from great distances and can detect blood in concentrations of much less than one part blood per million parts of sea water. They will follow the odor of a bait up current or even along a wind-driven surface film on the water. In following this trail, they swim a sinuous course, constantly testing for the area of strongest concentration of odor, much as a hunting dog does in trailing a rabbit. The olfactory stimulus appears to induce exploratory behavior, but other sensory stimuli, such as electromagnetic fields or tactile stimulation may be needed to initiate feeding or snapping at the food source.

Although some sharks seem to eat almost any available food, sharks do have a fine sense of taste. When biting certain organisms such as the Moses sole, a fish that secretes a toxic substance (pardaxin), the whitetip reef shark is repelled and swims away violently with its mouth agape. Taste buds are present in the mouth and pharnyx of sharks and are probably responsible for the shark's final acceptance or rejection of prey items. Certain species of sharks with nasal barbels may use them to taste substances, as has been demonstrated for many bony fishes with barbels (plates 11c and 12b). The mandarin dogfish, *Cirrhigaleus barbifer*, of the western Pacific, for example, has very long barbels, whose function has not been tested, but that could be used to trail over the sea bottom to sense prey by taste. Barbels, however, might also function as organs of touch.

Is sight important in detecting prey? Yes, especially at distances of 15 m or less. The visual system of sharks is sophisticated and complex, despite their reputation for having poor eyesight, especially in daylight. Studies of over 20 different species reveal that most have duplex retinas, that is, composed of two types of light receivers (called photoreceptors): rods and cones. In man, rods function in dim light and form colorless sensations (shades of black and white) and cones function in bright light and detect color. Theoretically, therefore, sharks are able to use their vision in both nighttime and daytime activities, and may also possess color vision. Only one shark, *Hexanchus griseus,* the sixgill, is known to have only rod receptors and, therefore, should lack color vision. Laboratory experiments have demonstrated that some sharks can be taught to distinguish between vertical and horizontal stripes, upright and upside down triangles, and various shapes. Other studies indicate they might be able to choose between white and red targets; however, there it is possible that the sharks were responding to slight differences in light intensity rather than to color. Whether sharks discern colors such as red, blue, purple, green, yellow, and brown as do humans, is unknown. It is difficult to determine the existence of color perception in any organism, and sharks will require particularly ingenious experiments to divulge this capacity.

Although sharks probably cannot differentiate shapes or colors very well, they are particularly sensitive to light, movements, and contrast. Some species have fixed upper and lower eyelids, others have moveable lids, and sharks in four families have a third eyelid, or nictitating membrane (figure 14a). The lids appear to be protective devices rather than mechanisms to reduce the amount of light entering the eye. Movement of the nictitating membrane has been observed during feeding, when it may be pulled up over the eye as the shark bites, or when a strange object approaches or is passed. The presence of a tapetum lucidum and pupils that can dilate and constrict quickly, indicate that the eyes of sharks adjust to changes in light intensity.

In experiments with the lemon shark to determine rate and range of dark adaptation, the shark was tested to determine if it could see a flashing light. Lemon sharks normally do not use their third eyelid (nictitating membrane) to blink in response to light, but will blink in response to low-level electric shocks. A test shark was conditioned to blink in response to light by first subjecting it to an electric shock paired simultaneously with a flashing light. Ultimately the shock was eliminated and the shark, like Pavlov's dog, responded by blinking to the light alone. Having conditioned the shark to react to light, it was then possible to determine how much light was necessary to produce a blink. Using the same test apparatus, it was found that the dark-adapted eye of the shark reacted to ten times less light than did the eye of humans.

Some sharks hunt in moonlight, or at least very dim light, and combined with their other senses, see with enough acuity to follow and capture live prey. Although laboratory studies have given us a better idea of the visual capacity of sharks, it is not clear how sharks use this capacity in the wild. Furthermore, of the approximately 368 living species of sharks, experimentation has involved only a few, and generalizations derived from them are open to question.

What other means do sharks use to detect prey? Sharks have a unique and highly specialized bioelectrical sensory system scattered over the front of their heads that not only helps them detect and home in on prey, but may help them navigate relative to ocean currents and the earth's magnetic field. This system consists of several lateral-line canals and hundreds of skin pores, called the ampullae of Lorenzini (plate 10c), honoring the seventeenth-century anatomist who first described them. The ampullae of Lorenzini contain receptor cells that not only detect the presence of an electric field, but also the direction of its intensity.

Because all living animals, including those in the sea, produce electric fields, sharks can find their prey, at least at close range, without seeing or smelling them. They can sense stingrays and flounder buried in the sand and strike with great accuracy. Over a frequency range from 0 hertz to about 8 hertz (cycles per second), much lower than the human ear can hear, the smooth dogfish can detect a change in direction of intensity of 5 billionths of a volt per centimeter, the highest degree of electrical sensitivity known in the animal kingdom, better even than a good hi-fi stereo system. Odor cues may attract sharks at some distance, but at short range the electric fields emitted by prey are the strongest stimuli.

Much of the research on electric field detection in sharks was done by Dr. A. J. Kalmijn and his associates at Woods Hole Oceanographic Institution in Massachusetts, and is continuing at the Scripps Institution of Oceanography in California. Their experiments with dogfish and blue sharks have shown that not only can sharks detect the weak electrical fields of prey, but also may be able to navigate and orient themselves with respect to the earth's magnetic field, much as if they had an innate compass (figure 33). When sharks swim through the earth's mag-

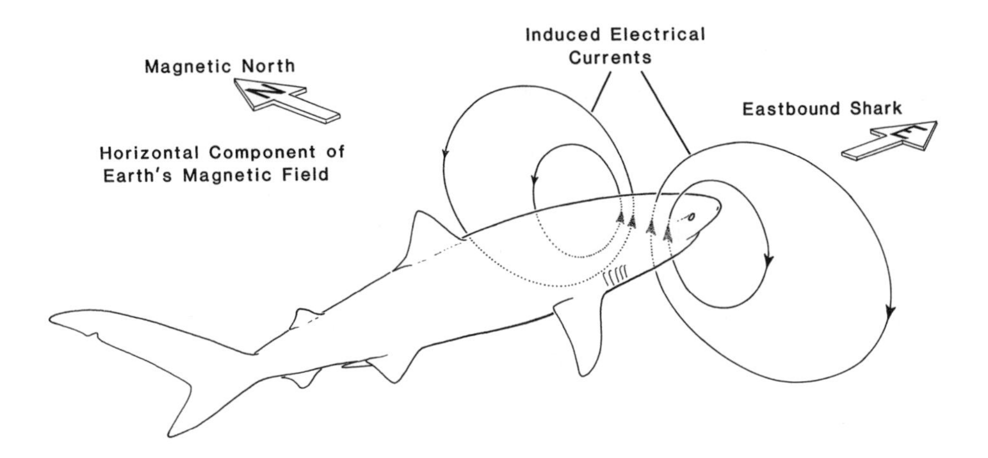

Figure 33. A shark swimming through the earth's magnetic field induces electric fields that provide the animal with the physical basis of an electromagnetic compass sense. (After Kalmijn, 1978)

netic field, changes are induced in their own bioelectric fields. By this means, sharks seem to be able to tell their direction relative to the north-south horizontal component of the earth's magnetic field and may be able to determine their latitude using the vertical component of the field. Sharks may also be sensitive to the weak electric fields induced by ocean currents flowing through the earth's magnetic field, and thus use them in navigation.

Further readings

Boord, R. L., and C. B. G. Campbell, 1977; Cohen, J. L., 1981; Gilbert, P. W., 1963; Graeber, R. C., 1978; Gruber, S. H., 1967, 1977; Gruber, S. H., and A. A. Myrberg, Jr., 1977; Gruber, S. H., and J. L. Cohen, 1978; Gruber, S. H., and P. Morrissey, 1978; Hodgson, E. S., and R. F. Mathewson, 1978; Kalmijn, A. J., 1971, 1977, 1978, 1982; Moss, S. A.,1984; Myrberg, A. A., Jr., 1978; Myrberg, A. A., Jr., et al., 1976; Ryan, P. R., 1981; Tester, A. L., 1963.

HOW SMART ARE SHARKS?

Smarter than one might think, but it would be a mistake to measure their mental abilities against those of humans. The term "intelligence" has many implications and definitions. Some scientists correlate intelligence with the ability to communicate complex thoughts, others with self-awareness, still others with adaptive

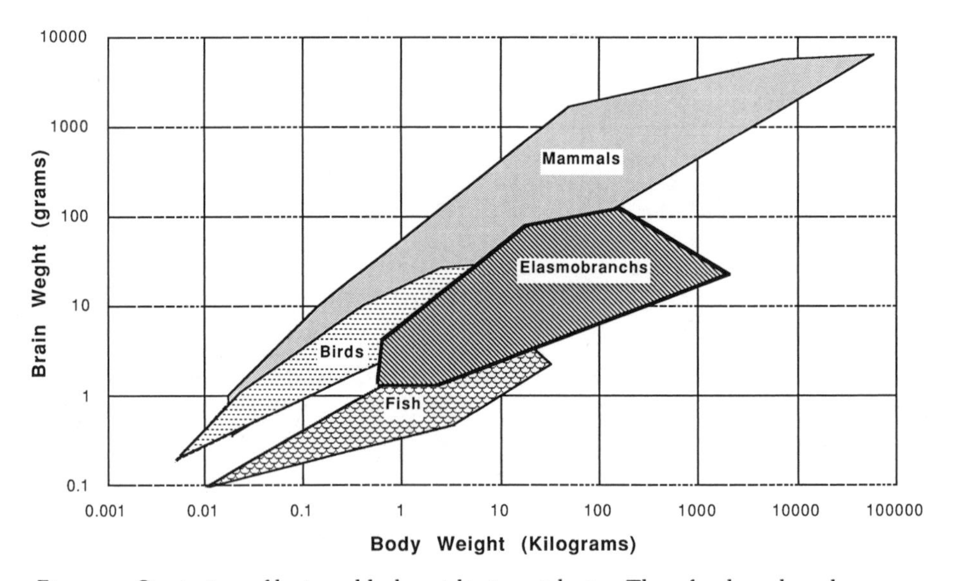

Figure 34. Comparison of brain and body weights in vertebrates. Those for elasmobranchs (sharks, skates, rays, and chimaeras) are encompassed by the striped area. (After Northcutt, 1977, and Moss, 1984)

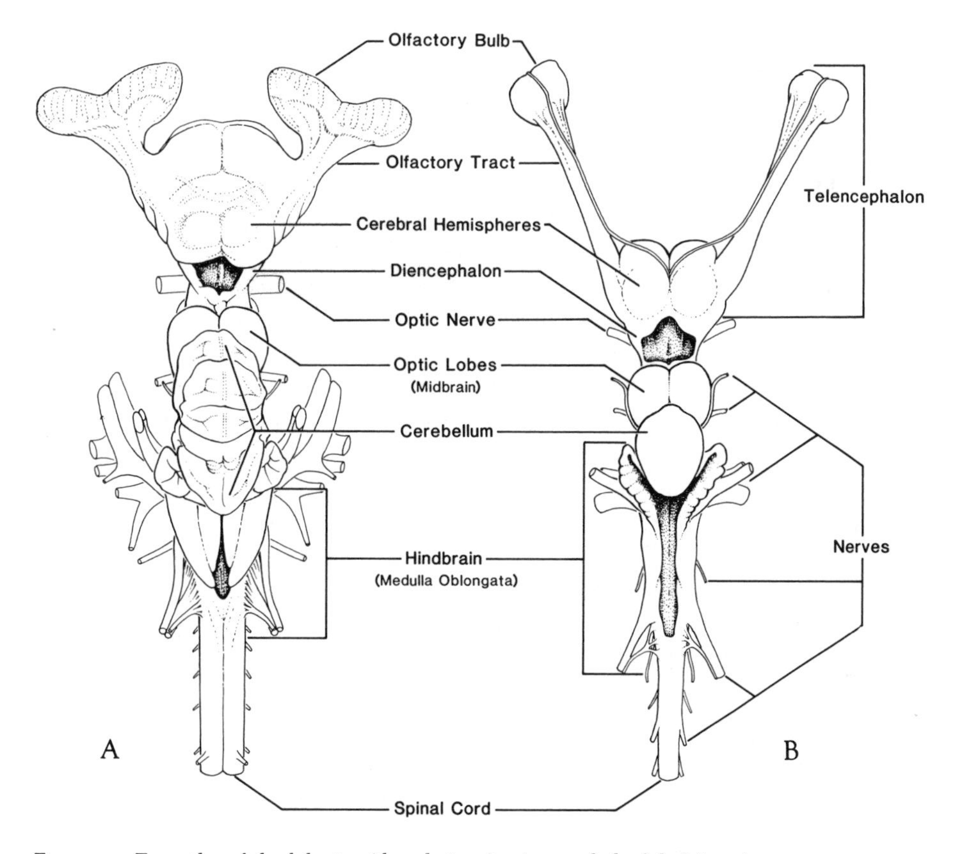

Figure 35. Examples of shark brains (dorsal views): a) smooth dogfish, Mustelus canis; *b) sevengill shark,* Notorynchus cepedianus. *(Modified from Northcutt, 1978)*

strategies requiring foresight and ability to learn. Because of the difficulties in capturing sharks and maintaining them in captivity, or conducting experiments with sharks living freely in the wild, we still know little about whether specific behaviors of sharks indicate intelligence or instinct.

Traditionally, large brains relative to body weight have been thought to correlate with complex behavior, such as is used in food manipulation and complicated exploration. This correlation holds for bony fishes and reptiles. Some species of sharks have relatively small brains, but many have large brains with brain-to-body ratios well within the ranges for birds and mammals, which have the largest brains and exhibit the most complex behavior (figure 34).

The shark brain (figure 35) consists of the hindbrain (or medulla oblongata), the cerebellum, the midbrain (or mesencephalon), and the forebrain (or telencephalon), which includes the diencephalon. The diencephalon contains the thalamus, epithalamus, and hypothalamus. Most of the several parts of the brain perform a sensory role. At the anterior end of the forebrain are a pair of cerebral hemispheres, each of which connects by a stalk to an olfactory bulb. The cerebral hemispheres make up over half the brain and are thought to integrate olfactory information. The midbrain is believed to integrate visual information and the cerebellum is believed to control muscular coordination, but there are still large gaps in our knowledge of the neurobiology of sharks. For example, it was long believed that a major portion of the forebrain was used in olfaction, but recently it has been determined that only 10 percent of the forebrain is so employed. The function of the other 90 percent is still a mystery.

Intelligence is most often equated with the ability to learn. We know sharks can learn at the same rate as do birds and rats. Lemon sharks are surprisingly good at learning tasks, and can learn a maze as well as a rabbit.

Experiments have demonstrated that sharks can modify their behavior if their correct responses are rewarded. Lemon sharks were conditioned to push a target with their snout and ring a bell and then obtain a reward of food at another location. They eventually learned to associate the mere ringing of the bell with the availability of food. Even more dramatically, two nurse sharks at the Naval Ocean Systems Center in San Diego, California, were trained to work rings over their heads, like those used in dolphin and seal shows, and swim to their trainer to receive food. It took 6 months to condition the sharks to do this, but they remembered how, even though they were being used in other behavior studies at the same time. Obviously, sharks can be trained to do things completely foreign to their natural instincts, but it is not clear how operant conditioning (the correct-response-reward technique used to train them) mirrors their behavior in nature.

How do sharks select particular stimuli to respond to from the myriad competing for their attention in their natural surroundings? We just do not know.

Further readings

Brownlee, S., 1985; Clark, E., 1959; Johnson, C. S., et al., 1984; McManus, M. E., et al., 1984; Moss, S. A., 1984; Northcutt, R. G., 1977, 1978.

WHAT DO SHARKS EAT?

Sharks are flesh eaters, feeding mainly on crustaceans, fishes, and mollusks. Some, perhaps most, are generalists, eating whatever food is available when they are hungry, which includes not only fishes and invertebrates, but marine turtles, birds, and even such unlikely objects as tin cans. If sharks eat an object they cannot digest, such as a turtle shell or bottle, they can regurgitate it by thrusting their stomachs up and out through their mouths to expel the object, then retracting the stomach. Such behavior, although known to occur, may be infrequent and only provoked under unusual circumstances.

Many species of sharks are selective about their prey and even seem to have preferences. Where a predominance of a certain kind of prey is consistently found among the stomach contents of a species throughout its range, it can be assumed

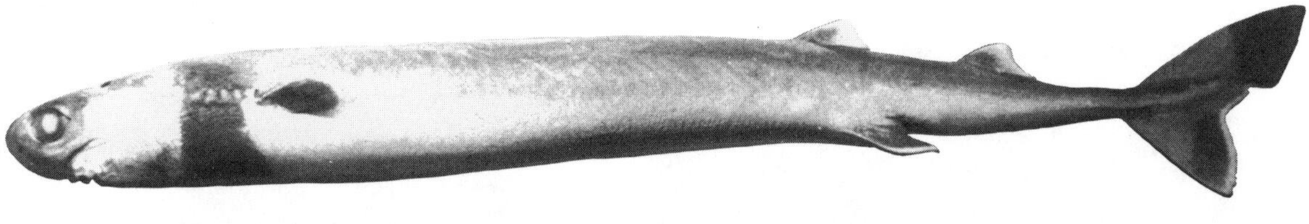

Figure 36. Cookie-cutter shark, Isistius brasiliensis, *Family Squalidae, circumglobally distributed in open ocean waters, primarily in tropical latitudes. The cookie-cutter shark derives its name from the round, cookie-like plugs it bites from its prey. (Paul Allen)*

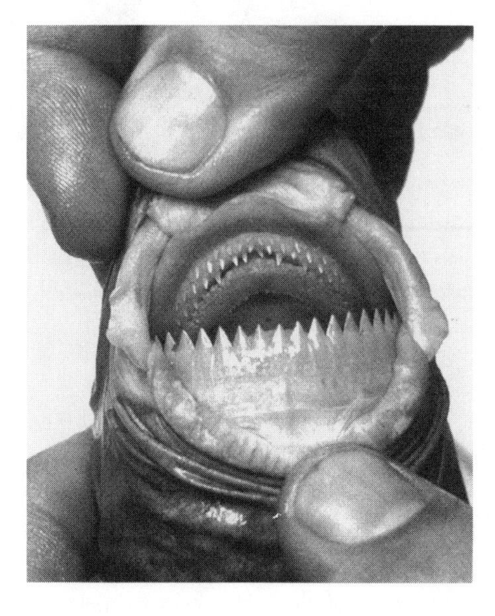

Figure 37. Mouth of the cookie-cutter shark. (C. Scott Johnson)

the shark prefers this kind of food. For instance, hammerheads prefer stingrays and horn sharks prefer sea urchins. The zebra shark prefers mollusks, although not to the exclusion of crabs, shrimps, and small bony fishes. The Greenland shark feeds primarily on bony fishes, but also eats marine mammals. It preys on live seals but will scavenge, devouring all manner of carrion from whaling and sealing stations. Parts of drowned horses, an entire reindeer, as well as sea birds, sea urchins, and jellyfishes have been found in the stomachs of Greenland sharks. All these sharks devour relatively large organisms, but the basking shark, one of the largest of all sharks, dines only on tiny plankton.

Yet one should sometimes be skeptical about reports of a species of shark preferring one prey over another. For example, there are reports that the shortfin mako shark prefers bluefish (*Pomatomus saltatrix*). The shortfin mako, however, is found throughout all warm seas, but the bluefish has a more limited distribution. It is common in the Atlantic, for example, but in the entire Pacific occurs only along the southeastern coast of Australia. Examined stomachs of shortfin makos in the northwestern Atlantic have contained mostly bluefish, both by weight and volume, probably because bluefish are so abundant there. What most sharks eat depends on what is available and how available (abundant) it is.

The main factor that dictates the prey preference of any shark is the kind of teeth it has: knife-like, spike-like, pavement-like, prickle-like, or so small as to be

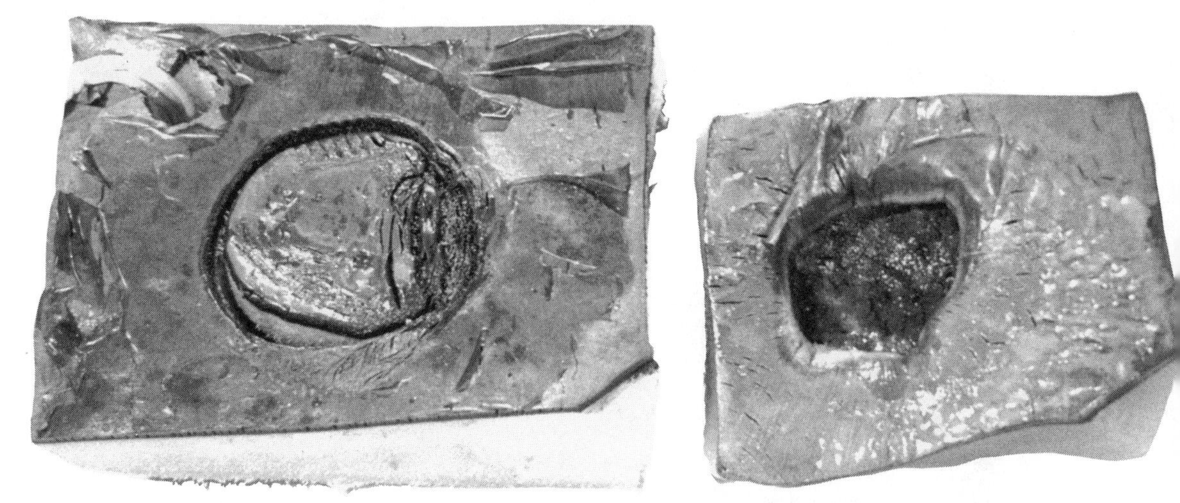

Figure 38. Pieces of whale skin showing (left) an aborted bite and (right) a healing wound made by a cookie-cutter shark. (T. Britt Griswold)

virtually absent. Sharks with formidable teeth, powerful jaws, and an upper jaw that can protrude are well equipped to gouge out huge pieces of flesh from their prey. They need not rely only on small fish for dinner but can feed on organisms much larger than themselves. For example, the cookie-cutter shark, one of the smaller shark species, takes its name from the plugs of flesh it carves from large whales (figures 36, 37, and 38).

The toughness of the prey does not appear to be a deterrent to feeding by some sharks. Even submarines have been bitten. Fiber optic communication cables lying on the ocean bed a mile deep have been bitten so severely that service was interrupted (electromagnetic fields radiating from the cable casing may be provoking the attacks).

A shark's bite exerts considerable force, which has been measured by an instrument called a gnathodynamometer (literally, jaw power meter). The maximum force so far recorded as being applied to a single tooth (with a 2 mm^2 tip; of a shark 2 m in length) is 60 kg (132 lb). This force converts to a tooth-tip pressure of 3 metric tons per square centimeter (= 42,674 lb per square inch). Because of profound differences between the jaw and tooth structures of humans and sharks, it is not possible to make an exact comparison of the total biting forces each can develop. Surprising as it may seem, tooth pressures developed by a human may exceed those reported for sharks. Based on static force measurement, the average forced biting force in an adult male human is 45–68 kg (forces as great as 159 kg have been recorded for Eskimos). *The Dental Science Handbook* published by the American Dental Association for 1970 gives biting pressures by humans as great as 30,000 lb per square inch, not greatly different from that of a shark.

The reason a shark's bite is so much more damaging than a human's bite is that the shark's teeth are razor sharp, whereas the human tooth is blunt, and more teeth are involved in a shark's bite than are involved in a human's bite. These differences combine with the shark's ability to project its jaws and latch on to its prey in ways a human cannot duplicate.

Further readings

Heemstra, P. C., 1980; Ramfjord, S. P., and M. M. Ash, Jr., 1971; Snelson, F. F., Jr., et al., 1984; Snodgrass, J. M., and P. W. Gilbert, 1967; Stillwell, C. E., and N. E. Kohler, 1982.

WHAT IS A FEEDING FRENZY?

From time to time, the presence of assorted and numerous stimuli in the water, such as an abundance of prey, blood, and irregular movements like those of a struggling fish, can trigger a type of behavior known as a "feeding frenzy." Frenzied sharks attack anything that moves—snapping, slashing, and biting even their own kind, without inhibition. The attack intensifies and becomes more ferocious as more sharks join the group. It is not known exactly why this spectacular behavior occurs, but possibly the sharks' brains become so overloaded with sensory cues that they lose all inhibition. Similar though not identical behavior occurs among other vertebrates and is common in birds. Different species of birds will "mob" or mass together to attack a common enemy such as a hawk, owl, or snake. Barn owls have been attacked and killed by herring gulls and snowy owls have been mobbed by arctic terns. Even human behavior can take on the appearance of a frenzy, as for example, in mob riots. Because sharks continue the slaughter, even when they themselves are mutilated, it has been asked

DO SHARKS FEEL PAIN?

Little is known about how the brain and nerves sense pain in humans, and far less is known about it in sharks. Since before the time of Aristotle, the nature of pain has been debated. Is it a physiological and/or emotional response? Human pain thresholds, which differ from one person to another, are regulated by chemicals produced in the brain. Some people are capable of receiving painful stimuli that would be intolerable to others. Sharks also may be able to receive painful stimuli yet be insensible to it. Blue sharks in a frenzy have been seen to return repeatedly to feed on whale carcasses, even though the sharks were fatally injured or badly mutilated by the whalemen fending them off. But at present there is no way we can measure, confirm, or deny the degree of pain sharks may feel.

DO SHARKS HAVE ENEMIES?

Some sharks prey on other sharks, but it is people who are sharks' greatest enemies. Sharks are among the most successful predators of the ocean. They prey upon most marine organisms, but few organisms prey upon them. Stomachs of large bony fishes and killer whales occasionally yield the remains of small sharks, but much more often, large sharks will eat smaller ones, and smaller ones may even bite chunks out of larger species. Some sharks do not hesitate to eat their own young. For that reason, possibly, females of some species do not eat while they are in the nursery grounds during the pupping season.

If sharks are not their own worst enemies, humans must be. Big game-fishing tournaments and commercial fishing operations for shark meat and hides have greatly affected shark populations in various areas. According to statistics accumulated by the National Marine Fisheries Service, in 1985, about 3.4 million pounds of sharks were caught by California commercial fisheries and sold to markets and restaurants. This contrasts with the 1975 catch of little more than half a million pounds. The amounts may seem insignificant when compared with the tuna fishery, which comprises over 90 percent of the commercial fishing in California and which landed over 61 million pounds in 1985. Sharks, however, have a relatively slow growth rate, often taking years to reach maturity, and then giving birth, in most species, to only a few young, so it may not take much fishing to have a disproportionately large depletion effect on the shark populations.

A survey of recreational fishing in the western central and northwest Atlantic Ocean and Gulf of Mexico estimated that 2,600 tons of pelagic sharks were caught in 1965. The catch increased to about 8,700 tons per year from 1969 to 1978, and to about 15,900 tons in 1980. Another survey along the Atlantic coast estimated that over 230,000 large sharks, such as the blue, dusky, lemon, shortfin mako, thresher (about 10,277 tons or 22.6 million lb), were caught by offshore fishing boats in 1978. Because the survey did not include dogfishes or sharks caught from the shore, the estimates are conservative. Both surveys indicate that increased pressure on these large sharks is to be expected in the near future. The June 1983 Cape May Shark Tournament, on the U.S. Atlantic coast, alone accounted for the elimination of over two hundred blue sharks, and in New Jersey and New York there may be as many as one or two fishing tournaments a week. Humans are the most formidable enemy that sharks will ever have.

Less threatening are the parasites that afflict sharks. Leeches, trematodes (parasitic flatworms), and copepods (tiny crustaceans, some species known as fish lice) may attach themselves to the body surface, on the gills, and in the mouth or nasal cavity. Various species of copepods parasitize preferred hosts. Blue sharks characteristically are infested with clusters of the round, black copepod, *Pandarus satyrus* (plate 7b), and the shortfin mako is host to *Dinemoura latifolia,* copepods that look like masses of shiny brown buttons. Tapeworms, some very long, frequently numbering in the thousands in a single shark, and roundworms are often found in the stomach and intestinal tracts of sharks. These shark-parasite associations may have little effect on the longevity of many sharks, but if the parasites are present in large enough numbers to cause severe damage to body tissues, the shark could die.

Further readings
Anderson, E. D., 1985; Benz, G. W., 1985; Casey, J. G., and C. E. Stillwell, 1983; Casey, J. G., and J. J. Hoey, 1985; Grimes, D. J., et al., 1985; Klausewitz, W., and Hjalmar Thiel, 1982; Springer, S., 1960.

DO SHARKS GET SICK?

Yes. An autopsy of a sand tiger that died at the National Aquarium in Baltimore, for example, revealed that it was parasitized by tapeworms and roundworms, had liver disease, two possible tumors, and signs of meningitis. Tumors in sharks are believed to be rare, although the subject has been little studied. Research on fish diseases has focused on economically more important marine species than sharks, so scant attention has been paid to their diseases. That few sharks have been found with tumors or active diseases may be due to their highly predatory and cannibalistic nature. If a shark is sick or gives evidence of weakness, it may well be eaten by another shark.

Staff members of the Smithsonian's *Registry of Tumors in Lower Animals* conducted a computer search for information about tumors in fishes. Their search covered data about tumors extracted from about 4,500 published reports and unpublished information on about 3,000 museum specimens. Among this large data base were only 30 reports of tumors in sharks, of which only 10 or 20 reports are considered well documented. The others are probably based on incorrect diagnoses. A few valid examples are the thyroid tumors reported in a dogfish, a tumor in a cat shark, and a leukemia type of tumor in a sandbar shark.

The stress of being captured, transported to, and held in an aquarium traumatizes some individuals or species of sharks, and causes them to refuse to eat and/or deplete their energy reserves. Bacterial (*Vibrio*) and parasitic (tapeworms, roundworms) diseases are common in sharks held in aquariums. Even though a healthy shark in the wild appears to handle these infections with little effect, weakened sharks in captivity succumb. Worms infesting the skin and causing hemorrhages are known to have caused the death of four out of five Galapagos sharks held in the Bermuda Aquarium since 1981. Septicemia (blood poisoning) killed a whale shark that had remained in captivity for an amazing 360 days in the Okinawa Expo Aquarium in Japan.

WHAT CHARACTERIZES THE MAJOR GROUPS (ORDERS) OF SHARKS?

Order Hexanchiformes—Cow Sharks and Frill Sharks

The hexanchiforms are a small group consisting mainly of deep-water or bottom-dwelling species living in temperate and tropical seas of the world. The order comprises two families, four genera, and five species: sharks primitive in form, that is, bearing resemblances to ancient, extinct species, and characterized by having 6 or 7 pairs of gill slits, one dorsal fin (without a spine in front), and an anal fin. The upper-jaw teeth are shaped differently from the lower-jaw teeth in hexanchoids. The upper teeth are fang-like and similar to those found in many modern species, whereas, in one family, Hexanchidae, the lower teeth are broad, slanting, and comb-like, and unlike those of most other living sharks. The sixgill, *Hexanchus griseus* (figure 39), and bigeyed sixgill, *Hexanchus vitulus*, are deep-water species that occur worldwide. The larger of the two, *H. griseus*, attains a length of at least

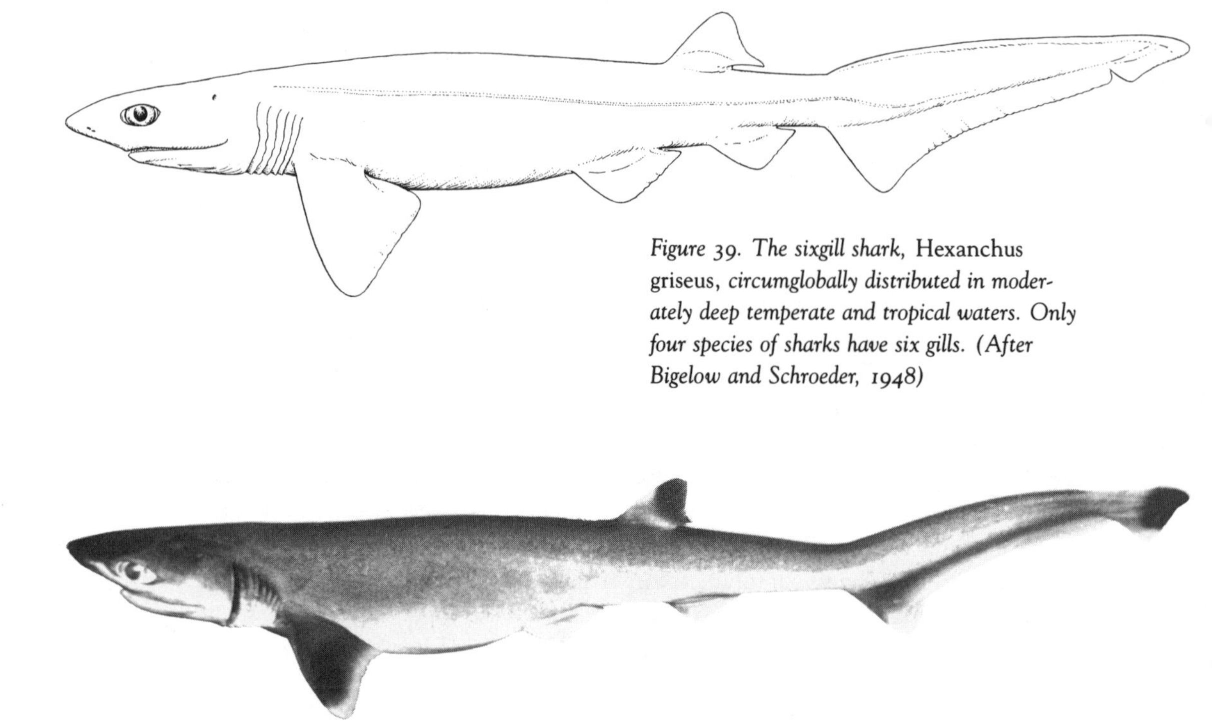

Figure 39. The sixgill shark, Hexanchus griseus, *circumglobally distributed in moderately deep temperate and tropical waters. Only four species of sharks have six gills. (After Bigelow and Schroeder, 1948)*

Figure 40. Sharpnose sevengill shark, Heptranchias perlo *(juvenile, 466 mm total length, off South Carolina), Family Hexanchidae, found in all shallow to moderately deep continental shelf waters, except notably absent from the entire central Pacific. (Don Flescher, National Marine Fisheries Service)*

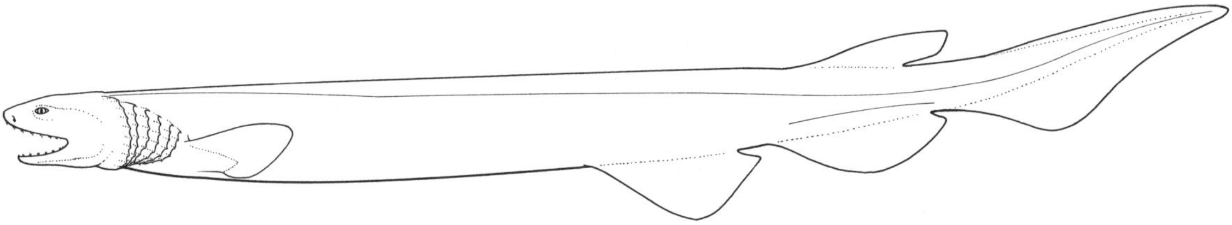

Figure 41. The rare frill shark, Chlamydoselachus anguineus, *deep-water inhabitant of cool-temperate seas (except the western Atlantic), has what appears to be a frilly collar about its throat, but which is the margin of the first gill opening running almost completely around the head. The frill shark attains a length of almost 2 m. (Modified from Cadenat and Blache, 1981)*

4.8 m. Both species have a broad head, thick body, long tail, and give birth to live young; 108 pups were counted in one litter. Other hexanchiforms are the "sevengills" belonging to the genera *Heptranchias* (figure 40) and *Notorynchus*, and the rather rare frill shark, *Chlamydoselachus* (figure 41), which has six gills.

Some tales of sea serpents may have been inspired by the eel-like frill shark, which has been recorded as reaching 1.92 m. An unlikely shark, this odd creature has a long snake-like mouth at the end of its head rather than an underslung one as occurs in most sharks. It sports what appears to be a frilled collar about its throat, but which is actually the margin of the first gill opening continuing almost completely around the head. Found mainly in deep water, this shark feeds primarily on octopuses and squid and is known to bear 8 to 12 young. Few frill sharks have been caught or observed, and much of the life history of the species remains a mystery.

Order Squaliformes—Dogfish Sharks

The Squaliformes is second among sharks only to the Carcharhiniformes (requiem sharks) in abundance and number of species. It comprises 3 families, 20 genera, and about 84 species of mainly deep-water inhabitants. The squaloids have a cylindrical or somewhat flattened body with two dorsal fins, but no anal fin. Many of the species have a spine anterior to each dorsal fin. The order includes the very large (average size, 3.4 m) Greenland shark that lives in the cold Arctic seas, and numerous small (less than 0.3 m in length) deep-sea species, with luminous organs.

Cookie-cutter Sharks

The most bizarre members of the Squaliformes are the cookie-cutter sharks, which belong to the genus *Isistius*. Though small in size, these sharks have relatively large teeth and powerful jaws. They frequently prey on fishes such as marlin and tuna, and attack even dolphins, elephant seals, and large whales. Even a U.S. nuclear submarine has been scarred by the cookie-cutter. These sharks attach themselves to their victims with lips modified for sucking, and carve out plugs or cones of flesh. *Isistius brasiliensis*, the common cookie-cutter, twists and rotates its body to remove flesh, leaving crater- or crescent-shaped wounds on its prey. Crater wounds found on cetaceans have been as large as 4 or 5 cm by 7 cm, and as small as 1.2 cm by 2 cm (figures 36, 37, 38). Lampreys may be responsible for some of the wounds, but the smaller diameters correspond well with bite widths for cookie-cutter sharks of from 14 cm to 50 cm in length.

Isistius plutodus, the largetooth cookie-cutter, although attaining a maximum total length of only 42 cm, is armed with very large triangular-cusped teeth, relatively the largest teeth of any shark species; in height, the teeth are approximately 25 percent of the head length. Though primarily a squid eater, *I. plutodus* also carves out long plugs of flesh from its prey. A plug of fish flesh taken from the stomach of *I. plutodus* was the same height and width as the diameter of the shark's mouth, but twice as long.

Isistius brasiliensis has been caught in water of depths from 85 m to 3,500 m in tropical and subtropical oceans. Because the sharks were caught by midwater trawling nets that passed through this range of depths, the exact depths of occurrence for the species are unknown. It may live at even greater depths and is thought to migrate vertically at night seeking prey closer to the surface. Like many other deep-ocean dwellers, it has large eyes, and on the undersurface of its body, luminous organs called photophores. *Isistius brasiliensis* is among the most luminescent of all sharks, the hundreds of photophores on the undersurface of its body and head emitting a ghostly greenish light. It is not known what purpose these photophores serve, but it is believed generally that photophores aid in attracting prey or in recognizing and attracting other members of the same species, which may be sparsely distributed in the deep, dark sea.

Dwarf Dogsharks

The Squaliformes include the smallest of all sharks. These are the two newly discovered deep-water dwarf dogsharks, *Etmopterus carteri* and *Etmopterus perryi* (plate 3a), discovered off the Caribbean coast of Colombia, the former with a maximum length of 21.2 cm (8.4 in), and the latter, 20 cm (7.9 in). Sexual maturity in *E. carteri* is attained in both males and females at a length of 18 cm. In *E. perryi* males are mature at 16 cm and females at 19 cm.

Only a little larger than the previous two species is *Squaliolus laticaudus* (plate 3b), the cigar shark, named for its physical appearance. The female grows to about 25 cm, but reaches sexual maturity when as small as 17 cm. The male reaches only 22 cm and may be mature at 15 cm. This tiny shark was originally reported from Japan, but subsequently was found in the North Atlantic, western South Atlantic, and the western Indian Ocean. During the day it remains at a depth of about 500 m, but at night, following its prey, migrates upward to 200 m. As is common with many of the deep-water species, this shark is dark brown in color, with no countershading of its underside. It possesses luminous organs, but they have never been observed functioning. Squid and other luminescent fishes such as lanternfishes are eaten by the cigar shark. Twelve mature eggs have been found in a single ovary of a female, but it is not known how many young are actually born.

Spiny Dogfish

Among the common refuse of university and college biological laboratories are the remains of dogfish sharks, *Squalus acanthias* (figure 42). Almost every anatomy or biology student who has wielded a scalpel is familiar with the internal organs and structure of this species. Although the spiny dogfish does not adjust well to captivity, it is plentiful and easily caught. As a result, it is widely studied by scientists and students, and more is known about its biology than that of most other sharks. Many of the advances in biomedical studies, particularly in physiology and immunology, are the result of research on the spiny dogfish. Unremarkable in appearance, the spiny dogfish is slate gray or brownish gray above and pale gray to white below. Usually, small pale spots are scattered over its body. It has sharp, blade-like teeth and weakly venomous fin spines.

The distribution of this species is peculiar. It is absent from the world's warm equatorial waters, but is present in the cooler waters on both sides. Such distributions are called bi-temperate, and how they are attained is a subject of controversy among biologists. One possible explanation hinges on past periods of the earth's history when the seas were cooler and the equatorial regions much narrower than at present. During such periods the species could have crossed the equator and become established on both sides. Later, when the climate became warmer and the equatorial waters expanded, the populations would be forced so far apart they would no longer be able to cross the tropics. Although there are no recognizable differences between populations on either side of the equatorial regions, it is possible that prolonged separation will result in the evolution of separate species.

Female and male spiny dogfishes vary in size. At maturity, the females are 70 to 124 cm, a bit larger than the males, 59 to 100 cm; however, a maximum size of 160 cm has been reported for this species. Spiny dogfish are calculated to grow more slowly in the Pacific than in the Atlantic but reach a greater size.

Similar to the method of counting the rings on the trunk of a tree, estimates of the age of spiny dogfish have been determined by counting the annual growth rings occurring on cross sections of its fin spines. Studies of spine growth and retrieval of tagged sharks indicate that spiny dogfish are slow growers and late in

Figure 42. Spiny dogfish, Squalus acanthias, *one of the most plentiful and well studied of all sharks. (Smithsonian Institution)*

maturing. Nonetheless, this species enjoys a ripe old age of at least 25 to 30 years, with some estimates of ages of females in the northeastern Atlantic of as much as 40 years and from the northeastern Pacific of more than 50 years.

Males, reportedly, are sexually mature at about 11 to 14 or more years of age, and females at from 10 to 35 years (depending on the validity of the research and the geographic location of the populations studied).

Spiny dogfish, like all squaliforms, are live bearers (ovoviviparous), with litters of from 1 to 20 pups. Mating probably occurs in the winter followed by a gestation period of from 18 to 24 months (plate 14a). In the northwestern Atlantic, most young are born from November to January, with reports from other areas indicating that most births occur in the cold months.

Of all the sharks found off the Atlantic and Pacific coasts of North America, the spiny dogfish is the most numerous and most important economically. Off the East Coast, it forms large migrating schools, often segregated by sex and size into aggregations of mature males, immature females, mature females, or small immature individuals of both sexes. Where, when, and at what age these sharks congregate and migrate are predictable, in part, based on extensive tagging studies made by commercial fishermen and scientists.

The seasonal movements of spiny dogfish appear to be correlated with water temperatures between 7 and 15 degrees Celsius (42 to 53 degrees Fahrenheit). They move inshore in the spring with the warming of the water and travel northward along the East Coast to Newfoundland and Labrador and south as far as Cuba. When the water becomes warmer than optimum in the south, they migrate to deeper waters, appearing again along the coast in autumn. Finally, when the temperature becomes colder than optimum, they move back to deep water.

Spiny dogfish feed on all kinds of small bony fishes, worms, squid, and other invertebrates, but will occasionally prey on species larger than themselves. The name "dogfish" was derived from their habit of assembling in large "packs" of thousands of individuals, and greedily attacking schools of prey species, such as cod and halibut. Commercial fishermen regard them as pernicious pests driving off or devouring valuable fishes and tangling and destroying fishing nets when caught in large numbers. They have been accused of seriously damaging the salmon populations in British Columbia. Although dogfish will feed on salmon trapped in a net, there is little evidence that they prey upon salmon naturally in the wild.

The only enemies the spiny dogfish has seem to be large sharks, seals, the killer whale, and man. They are used to produce the fish meal and oil found in cat and dog food, but are shunned as a food fish in North America. They have a better reputation as a food item in Europe and England, where they are sold as "flake" and "rock salmon." Although skates and rays are also used, the spiny dogfish is often the fish part of the English dish, "fish and chips."

Further readings

Al-Badri, M., and R. Lawson, 1985; Garrick, J. A. F., and S. Springer, 1964; Jones, E. C., 1971; Ketchen, K. S., 1986; Le Boeuf, B. J., et al., 1987; Nammack, M. F., et al., 1985; Parin, N. V., 1966; Seigel, J. A., 1978; Templeman, W., 1984.

Pristiophoriformes—Sawsharks

This small and unusual group of sharks comprises only one family, two genera, and five species. The group is distinguished by having an elongated blade-like snout studded with teeth on either edge and two long tentacles, called barbels, under the snout. The barbels possibly serve as taste sensors. Sawsharks (figure 6) may use their saw-like snouts to disable small fish or perhaps to stir up bottom sediment in search of prey. They have two dorsal fins, no anal fin, and five gill slits, except for one species with six. Because they resemble one another in body form and in having a long toothy snout, the sawsharks are often compared with their distant relatives, the shark-like rays known as the sawfishes. Features that distinguish the sawfishes are the location of the gill slits on the underside of the head and the lack of barbels. They are also larger, reaching up to 6 m, than the much smaller sawsharks, which range from 0.75 m to about 1.2 m. Sawsharks are strictly marine inhabitants, but sawfishes enter brackish and freshwater.

Sawsharks are bottom dwellers with a limited distribution in temperate and tropical waters of continental and insular shelves. In the Atlantic Ocean they are known only from deep water (640 m to 915 m) in the vicinity of Florida, the Bahamas, and Cuba. In the Pacific Ocean they are found only in the region between Japan and southern Australia, and in the Indian Ocean, only from southeastern Africa and western Australia. Outside the Atlantic, sawsharks may occur in depths as shallow as 37 m and as great as 430 m.

Even though some species of sawsharks are abundant (one species, the longnose sawshark, *Pristiophorus cirratus,* forms schools), their natural history is poorly known. Sawsharks are ovoviviparous and have litters of from 7 to 17 pups. The pups may be as large as 28 cm to 35 cm at birth.

In southern Australia and Japan, sawsharks are taken by bottom trawlers for human consumption. The flesh of sawsharks, particularly the longnose sawshark and the Japanese sawshark (*Pristiophorus japonicus*), is of high quality and makes excellent eating.

Further readings

Bass, A. J., et al., 1975c; Springer, S., and H. R. Bullis, Jr., 1960.

Order Squatiniformes—Angel Sharks

Thirteen species of angel sharks are in the single family (Squatinidae) and genus (*Squatina*) forming this order (plate 12c). Sometimes mistaken by the novice for skates or rays because of their greatly flattened appearance, the angel sharks (figure 7) have a broad, flat head and a body with large, wing-like pectoral fins, two dorsal fins, no anal fin, five gill slits, eyes on top of the head, and a terminal mouth. They have a unique tail fin among sharks; the lower lobe is longer than the upper lobe. The gill slits are partly on the sides of the head and partly on the lower or ventral surface, and the pectoral fins are not attached to the head. In the skates and rays, the gills are entirely on the lower surface of the head and the fins are fused to the head.

Angel sharks are sandy- or muddy-colored bottom dwellers that live in shallow coastal areas of temperate and tropical seas. They rarely grow longer than about 1.8 m. Perfectly camouflaged, they bury themselves in the sand with only their eyes visible and lie in ambush, ready to snatch and devour some small fish or crustacean with their needle-like teeth. They can quickly thrust out their jaws to catch a fish and can whip up their heads and snap quickly if annoyed or provoked.

Squatina squatina, the angel shark found off the continental shelves of Europe and North Africa, is nocturnal, swimming off the bottom at night and resting on the mud or sand during the day. It not only migrates vertically through the water, but migrates to the northernmost parts of its range in the summer. This shark is ovoviviparous, as are all of the angel sharks about which details of development are known. Litters contain from 9 to 20 pups, which are about 24 cm to 30 cm long.

Though they are relatively harmless to humans unless touched or harassed, caution should be exercised when handling or unhooking angel sharks from bait lines. Many species make tasty eating, and substantial fisheries for them exist in southern California and Peru. For human consumption they are salt-dried, or eaten fresh and fresh-frozen; they are also ground for fish meal and oil; and their hide is used for leather.

Further readings
Natanson, L. J., and G. M. Cailliet, 1986.

Order Heterodontiformes—Bullhead Sharks

Only one modern family, the Heterodontidae, with one genus and eight species, forms the order of bullhead, horn, or Port Jackson sharks (plate 2f). This group is characterized by having two dorsal fins, each with a spine at its anterior end, an anal fin, five gill slits, and varied, tile-like, mostly crushing teeth within each jaw (hetero = different; dont = teeth). The common name, horn shark, refers to the strong dorsal fin spines, and Port Jackson refers to an inlet in New South Wales, Australia, where one of the species was first described. Other living sharks, such as the spiny dogfish, that have dorsal-fin spines, lack an anal fin, and none have variably shaped crushing teeth.

Because the location and kind of teeth are similar to those found in certain ancient species of sharks, the heterodontids are often referred to as "living fossils." Each jaw has different kinds of teeth in the front and rear. The front teeth are small and sharp for capturing swimming prey and the rear ones are large, broad, and flat for crushing the shells of crustaceans, mollusks, and sea urchins (figure 12). Bullhead sharks seem to feed mostly at night.

Even compared with other unusual sharks, the bullheads look strange. They have distinctive, broad, pig-like faces, with flat snouts, knobby ridges above the eyes, and deep grooves that connect the nostrils with the mouth. They are small, most species not over 1 m in length, sluggish bottom feeders, and inhabit mostly shallow, warm-temperate waters of the Pacific and western Indian oceans, occasionally straggling into the tropics. Although bullhead sharks are common in eastern Australia and there are reports of fossils from New Zealand, only one living bullhead shark, belonging to an Australian species, has ever been taken in New Zealand. How this shallow-dwelling animal traversed the deep seas separating these two countries is something of a mystery. Possibly a bullhead shark egg case, with developing embryo, was torn loose from its attachment by a storm and carried by ocean currents to New Zealand, where the egg hatched after arrival.

Much of our knowledge of shark courtship and mating activities is based on observations of the horn shark, *Heterodontus francisci,* at the Steinhart Aquarium in San Francisco. It is one of the few shark species able to adapt well to captivity and has lived over 12 years in that aquarium. The horn shark is oviparous, laying its eggs in cone-shaped egg cases with distinctive spiral or screw-like flanges (figure 31). Each egg case contains a single egg or embryo. In captivity, one female laid two egg cases every 11 to 14 days over a period of 4 months. In nature, however, mating takes place in December or January, and the cases are laid only for about 3 months, from February to April. The female carries the cases in her mouth and places them in rock crevices, where the cases harden. Until the young hatch, some 7 to 9 months after the egg cases are deposited, the cases remain securely anchored, resisting strong ocean currents and waves.

Despite their strange appearance, bullhead sharks are not aggressive and pose no appreciable threat to man. More often, because of their sluggish nature, they are harassed by divers and may bite if provoked.

Further readings
Dempster, R. P., and E. S. Herald, 1961; Springer, V. G., 1982.

Order Orectolobiformes—Carpet Sharks

Zebra, carpet, and nurse sharks, colorful wobbegongs, and the gigantic whale shark are all members of this order (plate 2: a, b, c, and d; frontispiece; plate 11c; plate 12b). It seems absurd that these diverse species should be related, but indeed they are, illustrating how different in appearance related forms of life may be. The orectolobiforms comprise 5 families, 13 genera, and 31 species. Most of the species occur in shallow, tropical waters, primarily in the Indo-West Pacific, and especially along the coast of Australia. The anatomical features that relate the species are their pig-like snouts, similar to those of the bullheads, fleshy nasal barbels at the front of the mouth, grooves between the nostrils and the mouth, the lack of dorsal spines, the presence of small eyes above and behind the mouth, and spiracles behind the eyes. These sharks all have two dorsal fins, an anal fin, and five gill slits.

Except for the whale shark, most species of the order are bottom dwellers. Many of the species are adorned with colorful stripes and spots, perfect camouflage for hiding among rocks and weeds, in crevices, or buried in the sand. From these places of concealment, they ambush prey such as small fishes, shrimps, crabs, sea urchins, and other invertebrates.

Nurse Sharks

Only one species of orectolobiform, the nurse shark, *Ginglymostoma cirratum* (plate 2c), inhabits the shallow waters of the western Atlantic off North America. This is the shark most frequently encountered by divers off Florida and in Caribbean waters. It also occurs in the eastern Atlantic, from the Cape Verde Islands in the north to Gabon in the south, as well as the eastern Pacific, from southern Baja California (Mexico) to Peru. Because it is abundant and adjusts well in captivity, it is often exhibited in oceanariums and aquariums. Some individual nurse sharks have been maintained alive in aquariums for as long as 25 years. The nurse shark attains a length of up to 3 m, although most of the individuals encountered are much smaller. Growth rate is approximately 10 cm and 2.7 kg a year.

Unlike many other species of sharks, which must swim constantly to breathe, the nurse shark can lie quietly on the bottom and pump water over its gills. It is a drab-colored, sedentary, unaggressive species that allows divers to approach closely. Its apparent sluggishness belies its ferocity, however, if it is provoked. It has small but very sharp teeth and can inflict serious though not fatal wounds if prodded, pushed, or mishandled.

The nurse shark has been used for food, but it is more often valued for the leather made from its thick, tough hide.

Further readings

Carrier, J. C., 1987; Clark, E., 1963; Clark, E., and K. v. Schmitt, 1965; Randall, J. E., 1961.

Order Carcharhiniformes—Ground Sharks

The ground sharks comprise 8 families, 48 genera, and about 205 species. They are the most abundant sharks in terms of numbers of genera, species, and possibly individuals. These sharks are found primarily in tropical and temperate coastal waters, but some species are oceanic or enter freshwater.

The order consists of a variety of large and small species, most with an appearance that people imagine when they think of sharks (plate 4c). A few look more exotic, such as the hammerheads. Most carcharhiniforms have a wide mouth, a third eyelid or nictitating membrane, and sharp-edged teeth of more-or-less uniform shape within a species; the bases of the teeth in the outer row overlap, and frequently the largest teeth are on the sides of the jaws. All have two dorsal fins, no spines, an anal fin, and five gill slits. They eat a varied diet that commonly includes other sharks, rays, bony fishes, crustaceans, and cephalopods. Marine turtles and parts of seals have been found in the stomachs of members of the larger species.

The carcharhiniforms exhibit all three types of reproduction found in sharks, but the different types predominate in different families. For example, as far as is known, all the hammerheads, family Sphyrnidae, are viviparous, as are all the requiem sharks, family Carcharhinidae, except the tiger shark, which is ovoviparous. In the cat sharks, Scyliorhinidae, all those species for which the reproductive type is known are oviparous. In the houndsharks, Triakidae, about half the species are viviparous and half ovoviviparous.

Though most of the carcharhiniforms are less than 2 m in length when fully grown, many of the familiar, large, and dangerous species of requiem sharks attain much larger sizes. The maximum length for a bull shark, *Carcharhinus leucas*, is 3.2 m (over 10 ft), although the species rarely exceeds 2.8 m. The tiger shark, *Galeocerdo cuvier*, one of the largest species, is reliably reported to reach a length of 5.9 m (about 19 ft), with unverified records up to 9.1 m; and the maximum recorded length for a female dusky shark, *Carcharhinus obscurus*, is 3.6 m. By contrast, one of the smallest known sharks, *Eridacnis radcliffei*, a cat shark, ranging in size from 19 cm to 24 cm (7.6 in to 9.6 in), is also a member of the order.

Of all the sharks, the requiem sharks, family Carcharhinidae, most influence man's activities. Many of these sharks are of economic importance. They are large, voracious predators that often tangle and damage fishing lines and nets, competing with man for desirable seafood resources and causing economic losses in the swordfish, mackerel, and tuna fishing industries. They attack bait and feed on bony fishes, small sharks, rays, mollusks, and crustaceans. Many of the sharks in this family are caught and used for food (flesh and fins), oil (from the liver), leather (from the skin), and fish meal (entire animal except skin). A few like the tiger (plate 10a), bull (figure 26), Galapagos, oceanic whitetip (plate 4d), and blue sharks (plate 12a) are notoriously dangerous to man and are known for their attacks on divers, swimmers, and victims of sea disasters. It is from these relatively few species that the order and family derive their common name: requiem—a mass for the dead.

Often, requiem sharks are caught unintentionally by sport fishermen and are regarded as a nuisance. Conversely, some of the species are eagerly sought for the challenge they pose and are officially designated as game fish. The blue shark and the spinner fight strongly when hooked, and the spinner (figure 43), which may occur in schools, often leaps high out of the water with a spinning motion after taking the bait.

Figure 43. Spinner shark, Carcharhinus brevipinna, *2.0 m male, offshore coastal inhabitant of warm seas, but notably absent from the central and eastern Pacific. The common name is derived from the shark's habit of spinning around its long axis during its feeding runs through schools of fish, and then leaping out of the water. (Gregg Skomal, National Marine Fisheries Service)*

School Sharks

Intensive studies of the school shark (also called tope or soupfin), *Galeorhinus galeus* (figure 44) particularly populations off southern Australia, have provided information for one of the better-documented accounts of the life history of a carcharhinid.

School sharks are recorded from the cool-temperate waters of the eastern and southwestern Atlantic, Mediterranean, western Indian Ocean, and eastern and southwestern Pacific. They occur most often near or on the bottom, at depths from 2 m to 600 m. They are abundant, and they migrate in small schools of a hundred or so individuals, moving toward the poles in the summer and toward the equator, or in southern Australia, to deeper waters, in the winter. On average, the schools travel about 16 km (about 10 miles) a day, but can cover a distance of as much as 56 km a day.

Off Victoria, Tasmania, and South Australia, male school sharks reach maturity and mate when they are 8 to 9 years old. Females mature when about 10 years old and have given birth by the time they are 12. Size at maturity varies, but males are sexually mature by the time they reach 132 cm in length, and the smallest mature female recorded was 135 cm. Mating occurs from May to June in the deep waters of the edges of the continental shelf.

School sharks are ovoviviparous and carry their developing young only from June to December. This six-month term is one of the shortest gestation periods of any elasmobranch. The number of young is reported variously from 17 to 41 per litter and is directly related to the length of the female. One pregnant female, 144.5 cm long, carried 17 pups, whereas a larger female, 159 cm long, carried 41 pups, the greatest number recorded for the school shark in southern Australia. Eastern Pacific females, which attain a greater length, have from 6 to 52 pups in a litter.

In the months of November and December, gravid females enter the protective environment of shallow bays and estuaries to give birth, and then quickly head back to the offshore feeding grounds, leaving their newborn pups to fend for themselves. Most of the young will leave the nursery areas in the southern hemi-

Figure 44. School shark, Galeorhinus galeus, *aobut 1.3 m total length. (J. L. B. Smith Institute of Ichthyology)*

sphere summer (March) to feed offshore in schools. As juveniles, most will return to the place where they were born or to adjacent estuaries the following late spring/early summer months. Some may stay in the estuaries for as many as two years before joining schools of immature sharks offshore. About 8 percent more males than females are born, but by the time the sharks are two years old the percentages are reversed. The cause of the apparent greater mortality of males is unknown.

Times of breeding and pupping in the Northern Hemisphere are almost the opposite of those in the Southern Hemisphere because the seasons are reversed.

The habit of feeding in schools is the basis for the common name of this species. Schools consist predominantly of one sex or may consist predominantly of immatures or of adults of a limited size range.

School sharks are mainly bottom feeders but also feed on midwater organisms. They prey on almost any fish in their surroundings, but rarely on other elasmobranchs. Squid and octopuses are among their favored foods.

The average life expectancy for school sharks is unknown, but results of tagging studies indicate that some males live more than 41 years and some females more than 53 years. (For more information about the age of school sharks see the question, "How long do sharks live?")

School sharks are important to both commercial and sport fisheries. Since it began in 1927, the school shark commercial fishery in southern Australia has played an important role in the economic life of the country. These sharks make excellent eating, either fresh, frozen, or dried and salted, and their fins are in great demand by some Asian cultures for use in shark-fin soup. Because high levels of mercury have been found in Australian school sharks, legislation was passed setting a maximum permissible concentration of the metal in the sharks to be used for consumption. This action stopped the sale of large school sharks, which have the highest concentrations of mercury, but increased the fishing pressure on the immature sharks, causing concern that the species could become depleted.

Prior to 1940, a large fishery for school sharks existed in California. They were caught primarily for their livers, from which oil was extracted because of its high vitamin A content. The fishery declined because of overfishing, and the development of synthetic vitamin A lessened fishing pressure on the species as well. The lack of fishing pressure permitted the species to increase in abundance, although probably not to its former level. Because of an increase in the popularity of shark flesh as an inexpensive and tasty alternative to other fishes, school sharks in California waters are now sought again for human consumption. The California school shark fishery has revived to the point that overfishing may again become a problem.

Until recently, the populations of school sharks from the various oceans were considered to be different species. Currently, however, only one species of school shark, *Galeorhinus galeus*, is recognized.

Further readings
Heald, D., 1982; Olsen, A. M., 1984; Ripley, W. E., 1946.

Lemon Sharks

Because it grows to a large size, is abundant, and, particularly because it adapts well in captivity, the lemon shark, *Negaprion brevirostris* (plate 5d), another member of the Carcharhinidae, is widely used for laboratory and field studies. Lemon sharks do not need to swim to breathe and can be restrained and kept in experimental tanks for long periods of time without apparent ill effect. Studies in the wild have focused on age determination, growth rate, eating habits, and movements. Laboratory experiments have been devoted to the monitoring of caloric intake and the energy the animal expends to maintain itself, respiratory functions, food intake and digestion, and vision.

Much of the research on lemon sharks has been conducted by Dr. Samuel Gruber and his colleagues in the Bimini Lagoon, in the Bahamas, and the Florida Keys. They tagged 1,935 lemon sharks in the Florida Keys, and 280 at Bimini between 1979 and 1985 in an attempt to find out where they go, what they eat, and how they reproduce. Of these tagged sharks, 124 of the Florida sharks and 104 of the Bimini sharks had been recaptured by early 1986.

Lemon sharks grow much more slowly than previously believed (see the question, "How long do sharks live?"). Mating occurs in the spring and summer, with a gestation period of 10 to 12 months. In May or June pregnant females of the Bimini lagoon deposit their litters of 10 to 12 pups in shallow water less than 1 m deep, while males segregate themselves in the deeper waters off the lagoon. The birth of a shark in the wild is rarely seen and even less likely to be photographed, but Gruber's research team has succeeded in doing both. The pups emerge tail first, lie on the sand for about 10 seconds, strain against the umbilical cord, and finally pull free and swim away. The young remain in the shallow nursery areas for as long as 4 years, eating small fish, worms, shrimp, crabs, and octopuses, but growing only about 40 cm in length and 3.6 kg in weight during that time. Once they leave the shallows, they eat larger fishes, stingrays, lobsters and other crustaceans, mollusks, and occasionally, sea birds.

In one laboratory experiment, lemon sharks were given unlimited amounts of food twice a day for 100 days and allowed to feed until satiated. Feeding did not occur at the same rate at all times, but gradually increased and decreased over a 4-day cycle. The average shark (70 cm long) ate an amount equal to 3 percent of its body weight daily. By the end of the experiment the average shark's weight had increased 50 percent. Captive sharks on an unlimited diet grow four times faster than those in the wild. Other experiments using limited food offerings demonstrated that meal size was important in determining how often the sharks fed.

Gruber also tagged eight sharks with sonic tags, which emit high-frequency sounds that can be tracked by monitors on the lagoon floor. He was able to follow these sharks for periods up to 112 hours and pinpoint their location every 15 minutes. These sharks were active both day and night, with increased activity at dawn and twilight, moving strongly toward the sun at those times.

Gruber's intensive studies of lemon sharks continue and hold promise for providing the most detail on the behavior of a single shark species.

Further readings
Brownlee, S., 1985; Gruber, S. H., 1981b, 1985, 1988; Springer, S., 1950, 1963.

Swell Sharks

All sharks have some peculiarity that sets them apart from their fellows. For the swell sharks, it is their ability to blow themselves up like a balloon.

The swell sharks, genus *Cephaloscyllium* (plate 2e), are members of the cat shark family Scyliorhinidae. *Cephaloscyllium* comprises seven small species, variously limited in distribution to the shallow coasts extending from the eastern Indian Ocean to the eastern Pacific, but completely absent in the entire central Pacific Ocean. *Cephaloscyllium ventriosum* (plate 2e), the swell shark, occurs along the west coast of the Western Hemisphere from Monterey Bay, California, to southern Mexico, and along the coast of central Chile. It seems not to be on the coast from Mexico to Chile. The species prefers a rocky habitat with algae covering the bottom, and is usually found at depths of from 5 m to 37 m.

The swell shark is one of the larger cat sharks, reaching about a meter in length, with a broad flattened head, short snout, and large mouth with many small, sharp-pointed teeth. Brown blotches and numerous dark- and light-brown spots decorate the yellow-brown body and fins of the adult. Newly hatched swell sharks are lighter in color with well-defined markings.

The swell shark is oviparous, laying two eggs at a time. Each egg is contained in a large thick-walled capsule, greenish-amber in color (figure 30). The egg cases supposedly protect the eggs from predators during development, but one study found that the contents of 65 percent of the examined egg cases were missing, presumably having been preyed upon by snails. Depending on the water temperature, hatching takes place in 7.5 to 10 months. The newborn are 13 cm to 15 cm long with two rows of enlarged denticles down their back. The denticles, which point toward the tail, catch on the egg case and aid the shark in pushing itself out of the small egg case opening. These special denticles are shed soon after hatching.

During the day, the swell shark (sometimes several, stacked one on top of another) can be found resting in small crevices and caves, where it wedges itself into the rocks by swelling up. It inflates its stomach to an almost spherical shape by swallowing water or, if removed from the water, with air. At either end of the shark's cardiac stomach (anterior portion of the digestive tract) are sphincter muscles that contract when the shark traps water or air. To expel the water or air, the shark relaxes the sphincter at the anterior end of the cardiac stomach. The ability to inflate may be one way swell sharks protect themselves from large predators, and it clearly thwarts divers intent on capturing them.

At night the swell shark leaves its hiding place to feed on fishes and invertebrates. It has relatively large jaws for its size and many small pointed teeth that assist in holding prey that might seem too large for a small shark to eat. Sometimes the swell shark lies quietly waiting until a fish appears close by, then rapidly expands its mouth and gulps down its victim. At other times it slowly opens its mouth to catch an unwary fish that may swim into its jaws.

Although ordinarily a harmless species, if the swell shark is annoyed or harassed it may bite, but its bite is not particularly dangerous.

Further readings
Clark, E., 1947; Michael, S., 1986; Springer, S., 1979.

Order Lamniformes

The Lamniformes include some of the largest and most dangerous sharks as well as some large, harmless ones, and a few rare, bizarre forms (plate 6: a, b, and c; plate 9: b and c). The order comprises 7 families, 10 genera, and 16 species of extremely diverse forms ranging from the normal-appearing white shark with saw-edged triangular teeth, to the threshers with tails longer than their bodies and blade-like teeth, to the basking shark with a tail only a quarter of its body length, and minute hooked teeth. What characteristics bind together this strange assortment?

Oddly enough, there is no single characteristic by which one can recognize immediately that a shark is a lamniform. It appears that the uniting characteristics are a unique combination of structures that are variably present or absent in the other shark orders: a long mouth, conical or flattened snout, five gill slits, two dorsal fins, an anal fin, mouth situated behind the level of the anterior margins of the eyes, and no fin spines, no nictitating membrane, no barbels, and no deep grooves between the mouth and the nostrils. Most of the species have large front teeth, those in the upper jaw separated from the side teeth by a gap or by a row of smaller teeth.

Because the lamniforms exhibit no unique characteristic, one can wonder if the group is, indeed, a natural unit.

Most of the lamniforms attain a length of well over 4 meters. An exception is the crocodile shark, *Pseudocarcharias kamoharai,* which grows to a maximum total length of only 1.1 m (figure 45). The basking shark, second only to the whale shark in size, reaches a total length of at least 9.3 m to 9.8 m and possibly as much as

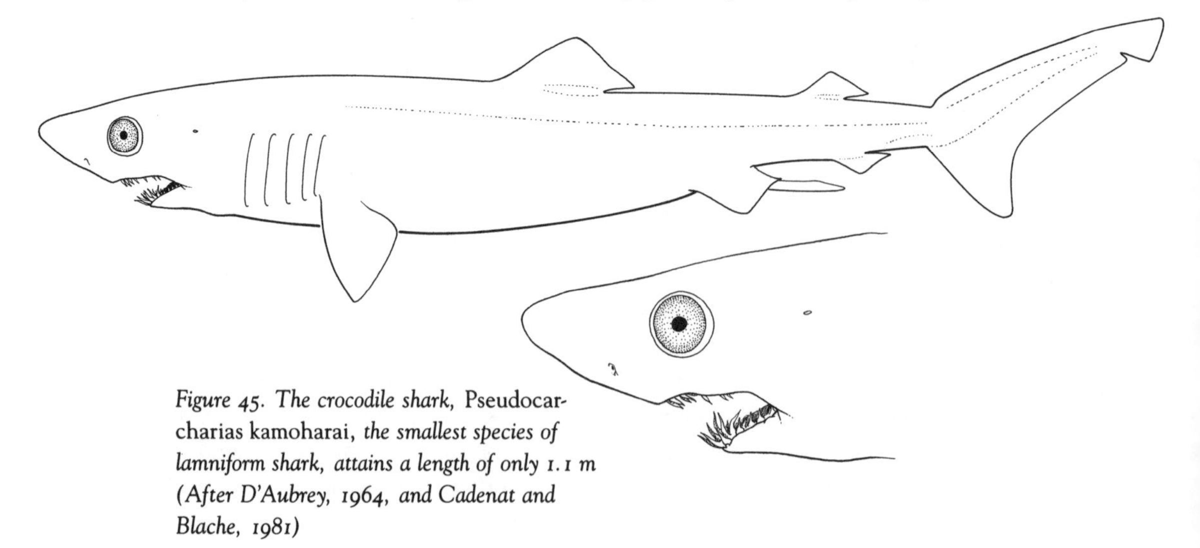

Figure 45. The crocodile shark, Pseudocarcharias kamoharai, *the smallest species of lamniform shark, attains a length of only 1.1 m (After D'Aubrey, 1964, and Cadenat and Blache, 1981)*

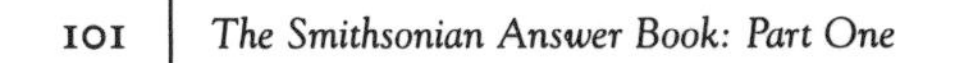

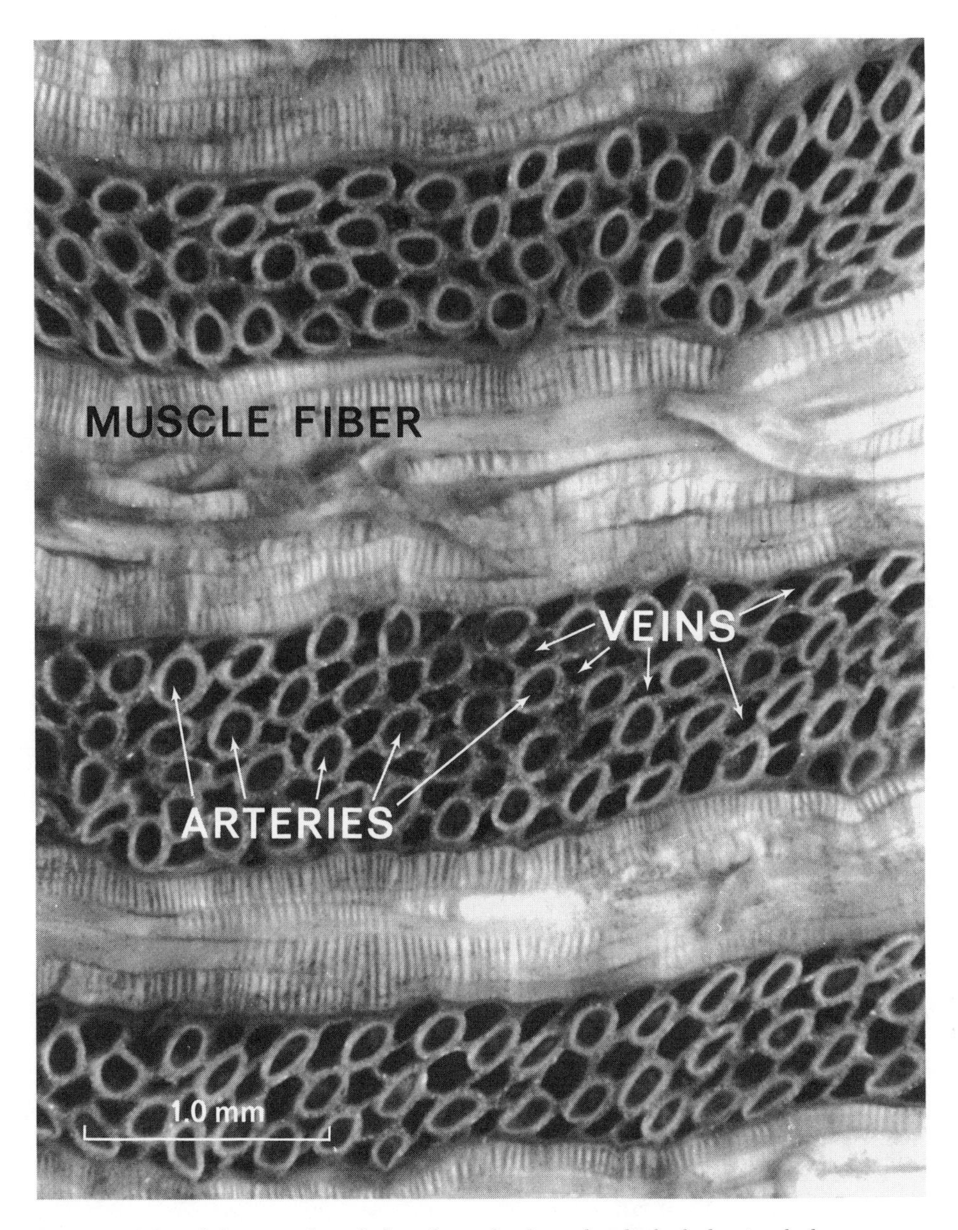

Figure 46. Magnified section through the red muscle of a porbeagle shark showing the heat exchanger (rete mirabile). Thick-walled, oblong appearing arteries surround thin-walled, irregularly shaped veins. Blood warmed by muscle activity is transported toward the heart by the veins. Cool blood carried from the gills by arteries is warmed by heat radiated from the veins. (F. G. Carey)

12.2 m to 15.2 m (verifying reported maximum lengths attained by the various species of sharks is like chasing the Loch Ness monster), although most individuals seldom are more than 9.8 m. The white shark attains a length of at least 6 m, and possibly as much as 8 m.

The lamniform sharks feed on a varied fare of small and large bony fishes, other sharks, crustaceans, squid, octopuses, and turtles. Some feed on schooling fishes like mackerel and herring, some on bottom species such as hake, and some, like the white shark, prey on marine mammals: porpoises, dolphins, seals, and sea lions.

Some lamniforms, because of their ability to conserve heat, are particularly well adapted to catching speedy prey like tunas and dolphins. Unlike other sharks and most fishes, lamniforms such as the mako, bigeye thresher, porbeagle, and white shark (plate 6: a, b, and c; plate 29b) are warm-bodied, maintaining temperatures in their muscles and viscera as much as 7 to 10 degrees Celsius above the surrounding water temperature. The increased temperatures are produced by a special circulatory modification called a *rete mirabile* ("wonderful net").

The rete mirabile (figure 46) is a tissue consisting of masses of fine intertwined arteries and veins that act as countercurrent heat exchangers. Arteries, carrying cool, oxygenated blood from the gills to the muscles and viscera branch out into smaller vessels and become intertwined with veins that are returning blood warmed by the activity of the muscles and other organs. As the vessels intermingle, the venous blood, delayed on its trip to the gills, diffuses its heat and warms the arterial blood bound for the muscles and viscera. This process permits the lamniforms to retain a greater portion of their metabolic heat than do most other fishes, in which the heat is lost to the environment by the blood returning more directly to the gills.

Because increases in temperature increase the speed of chemical reactions, it is believed that the elevated body temperature produced by the rete mirabile results in faster muscle contraction and faster rates of digestion, giving increased energy for swimming. Some tuna-like fishes also have retia mirabilia and higher body temperatures than the surrounding water.

Goblin Sharks

The most bizarre member of the Lamniformes is undoubtedly the goblin shark, *Mitsukurina owstoni* (plate 7a), which derives its common name from its unusual white to pale gray semitransparent skin. This grotesque creature (figure 47) with its

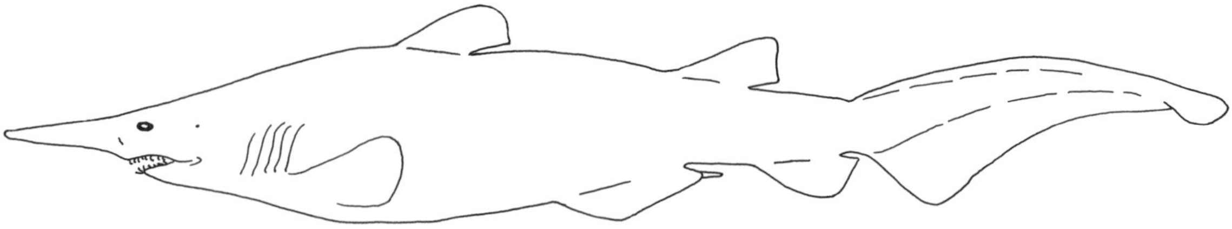

Figure 47. Goblin shark, Mitsukurina owstoni, *upper jaw retracted (compare with color plate 7a). (After Cadenat and Blache, 1981)*

soft, flabby body, overhanging, flat blade-like snout, protrusible jaws, and pointed teeth resembles, at least superficially, the fossil genus *Scapanorhynchus* of the upper Cretaceous more than any living species of shark. At one time *Scapanorhynchus* and *Mitsukurina* were believed to be the same species; however, recent studies have discovered that differences in the teeth and fins separate the two.

Because the goblin shark lives in deep water, at least 550 m deep, and is rarely caught, little is known of its biology and behavior. Only 36 goblin sharks are known to have been captured since the species was described in 1898. Of these, 23 are from Japan, the others from South Africa, Europe, Guiana, the Indian Ocean, New Zealand (not yet reported in the scientific literature), and Australia. Most were taken in water more than 350 m deep, although one or two reports are from a depth range of 160 m to 220 m. The presence of the goblin shark in Australia was confirmed only recently (1983) when a large, mature male was caught at 960 m. It weighed 212 kg and was 3.84 m long, the largest goblin shark ever recorded. This is one of the very few whole specimens to be collected, measured, dissected, and cast. The liver, an important buoyancy organ, weighed 9 percent of the shark's body weight; the heart 0.08 percent. The head, tail, fins, and claspers were removed and are preserved in the Australian Museum in Sydney.

How goblin sharks reproduce or what they eat is unknown. Stomachs of dead specimens have yielded few clues; most have been empty. The stomach of the recent Australian specimen was partially everted through its throat and contained what appeared to be a fish's swim bladder. No other food was found. Based on the kind of teeth the goblin has, it is believed they eat small fish and squid. The long snout is soft, covered with ampullae of Lorenzini, and seems to be used for detection of prey rather than as a device for digging.

Megamouth

Megamouth (plate 8: a and b) was a rare and startling discovery for ichthyologists when it was caught in 1976. Finding such a large, unknown sea creature by accident today is comparable to having a new kind of elephant amble into your own backyard. Megamouth (*Megachasma pelagios*) represented a new family, genus, and species. Even more amazing, and to the consternation of the ichthyological community, megamouth went without a scientific name until 1983, when the official description was finally published. This newest and rarest lamniform is known only from two mature male specimens: the first, not quite 4.5 m long, was caught at a depth of 165 m well south of the Hawaiian Islands, and the second, only a few centimeters longer, was caught at a depth of 38 m in 1984 off Santa Catalina Island, California.

The physical appearance of megamouth suggests that it is a weak swimmer and an atypical dweller of the depths. Unlike most deep-dwelling sharks, which are darkly colored on both surfaces, it is dark grey-blue above and silver-white below, a type of countershading found in near-surface inhabitants. It does, however, have loose skin and a poorly calcified skeleton, characteristics often found in deep-sea fishes. It is a filter feeder, similar in its eating habits to the whale and basking shark. Megamouth eats tiny euphausiid shrimp, copepods, and possibly, from the purple color of the liquid extracted from the intestines of the second specimen, pancake jellyfishes. When feeding, megamouth may migrate vertically up from deep waters, but in truth almost nothing is known of its movements.

Megamouth is named for its most striking feature, an extremely large mouth (plate 8b). The upper jaw and palate have a silvery, iridescent lining, which questionably luminesces, whereas the lower jaw and broad muscular tongue are black with the texture of velvet. One ichthyologist has suggested that when the shark opens its mouth, the black lower jaw becomes invisible in the darkness of the depths and the silvery lining of the upper jaw lures plankton like moths to a bright light. When the shark closes its mouth, sea water is retained by the tongue and palate and forced through the gill rakers, which strain out the tiny animals that megamouth eats. Wounds found on the body of the first specimen show it to be a victim of the cookie-cutter shark, making it the only known shark to be attacked by the cookie-cutter.

The second caught specimen of megamouth is on public display at the Natural History Museum of Los Angeles County, Los Angeles, California.

Further readings

Carey, F. G., and J. M. Teal, 1969; Carey, F. G., et al., 1981, 1982, 1985; Diamond, J. M., 1985; Lavenberg, R. J., and J. A. Seigel, 1985; Moss, S. A., 1984; Scholander, P. F., 1957; Stevens, J. D., and J. R. Paxton, 1985; Taylor, L. R., et al., 1983; Uyeno, T., et al., 1976; Wood, L., 1986.

PART TWO

The Supersharks

What is a supershark? All sharks are super, but here we spotlight those that are our favorites because of size, ferocity, or just because of an interesting problem a species presents.

Of all the sharks known and feared, none carries as notorious or unsavory a reputation as a voracious predator and maneater as does the white shark, *Carcharodon carcharias* (plates 6a and 11d). Called white pointer or white death by Australians because of its color, long, conical snout, and grim reputation, it is the largest of the flesh-eating sharks. Its belly is white, its back slate grey to black. Equipped with broad, triangular, serrated teeth (plate 10b), powerful jaws, a huge muscular body, and a metabolism that allows it to cruise slowly for long distances yet react with speed and agility to capture its prey, this remarkable creature is indeed a "supershark."

White Shark or Great White

The white shark is recorded from all the oceans, but is most abundant in the cool, temperate coastal and offshore waters of North America, southern Africa, and southern and western Australia. In the entire expanse of the central tropical Pacific, the white shark has been reported only from the Hawaiian Islands, where it is rare, and Bikini, Marshall Islands, where it is known from a single specimen. In the Indian Ocean, besides southern Africa and western Australia, it has been reported only from the Seychelles and, once, obscurely (Eibl-Eibesfeldt and Haas, 1959) and perhaps questionably, from the Red Sea. Despite considerable research activity on sharks in the Red Sea, recent Israeli publications (Darom and Baranes, 1980; Dor, 1984) on the fishes of that area state that the white shark is unknown there. Well-documented records of white sharks in the tropical Indo-Pacific should be brought to the attention of ichthyologists. Moderately large individuals are taken occasionally in tropical waters of the Atlantic and eastern Pacific, but newborn pups and smaller individuals appear to be restricted almost entirely to the temperate zones. Although the white shark may occur in offshore waters of great depth (up to 1,280 m), it also ventures close inshore to the surfline and enters shallow bays. The depth distribution of the species has not been established, but such records as exist would indicate that it normally occurs at depths less than 100 m, often in as little as 5 m.

Intensive ecological studies have been made on very few species of sharks and are notably lacking for the most dangerous ones. The data necessary to estimate abundance and size composition of shark populations are difficult and expensive to obtain. This is particularly true for the white shark because of its large size, mobility, and the infrequency with which it is encountered. We have no idea how many white sharks exist, how often they breed, how long they live, if they migrate and where, how much they eat, or how often they need to eat. Until 1986, when a 4.7 m female white shark with seven near-term embryos (100 cm to 110 cm in length) was captured off Japan, no well-documented records of gravid females existed. It is presumed the white shark is ovoviviparous and, as occurs in other lamniform sharks, oophagous: the developing embryos inside the mother feeding on the unfertilized eggs and cannibalizing one another until only the largest and most ferocious remains.

Much of the desired information about white sharks can only be obtained from tagging studies. The species is so uncommon, however, that of 11,863 sharks including 40 species tagged on the East Coast during the two-year period 1985–86 by the National Marine Fisheries Cooperative Shark Tagging Program, only one white shark was included. Of 51,379 tagged from 1963 to 1983, only 59 were white sharks. Furthermore, none of 1,634 tagged sharks that were recaptured during this same period was a white shark. The first tag return from a white shark was in 1984. The animal had been tagged off Long Island, New York, and was recaptured 2.5 years later, 88.7 km (55 miles) off Murrells Inlet, South Carolina.

The white shark does not appear to occur in large numbers and is uncommon compared with other species living in the same habitat. It seems to ply the waters alone or in pairs, rather than in schools. Small aggregations of about 10 individuals have been observed where food is abundant, such as around whales dead from stranding or near factory ships, or near expanding colonies of elephant seals, as may occur at the Farallon Islands near San Francisco. Feeding stations, where large quantities of blood, meat, or fish are regularly dumped in the water by avid photographers and fishermen, also attract groups of white sharks.

How fast does the white shark grow? How long does it live? White sharks seem to grow slowly and have a relatively long life span. Age and growth estimates based on vertebral ring counts correlated with body length from a limited sample of 21 white sharks from the eastern Pacific seem to indicate this, but many more specimens of both sexes and of different sizes and from various geographical areas are needed to validate these estimates. Slower growth is reported for the species in Atlantic waters, 20 cm per year, compared with estimates of 25 cm to 30 cm per year for younger, and 21.8 cm per year for older individuals from the Pacific. A 4.75 m (16 ft) white shark from the Atlantic is estimated to be 20 years old, whereas a white shark the same size from the eastern Pacific is calculated to be only 13 to 14 years old. Some scientists who have studied the white shark have estimated that it might attain an age of 25 years.

White sharks undoubtedly grow to more than 6 m in length, but like the "fish that got away," reliable measurements of such large creatures are nonexistent. The largest white shark previously thought to be reliably measured was a female taken off Cojimar, Cuba, in 1945. It was reported to be 6.4 m (21 ft) long and to weigh about 3,324 kg (about 7,300 lb). Photographs of this shark published in 1974 were reevaluated in 1987 and show conclusively that it was no more than 5 m long.

Recently a white shark definitely longer than 7.3 m (24 ft)—but impossible to measure exactly—was caught off southern Australia. Only the jaws were retained, the teeth of which were smaller than expected for such a large individual. An official report of this specimen has not been made but is anticipated soon.

Almost as large was another white shark, again caught off southern Australia, this time by commercial shark fisherman Vic Hislop and documented in a 1984 television film produced in Queensland. Hislop claimed the huge beast measured 6.3 m (20 ft 8 in) but was impossible to weigh. A steel cable testing 2 1/2 tons snapped when hauling the shark out of the water. Subsequent attempts to lift and weigh the carcass only succeeded in severing the tail.

Presently, the largest white shark reliably measured is one that was 6 m (19.5 ft) long, no weight given, caught off Western Australia. Very large individuals may weigh as much as 2,000 kg (4,800 lb). Although no birth has ever been observed, the newborn are thought to weigh between 16 and 27 kg (36 to 60 lb). Little is known of the size of females at first sexual maturity, although one caught with embryos was 4.7 m. But, in the Atlantic, males mature at between 3 and 4 m fork length (10 to 13 ft), which converts to a slightly greater total length.

The catching of a large shark, especially along the coasts of the United States, is usually a newsworthy happening. Any such capture of a shark more than 3 meters long sets the telephones ringing at the Smithsonian for days afterwards as newspaper, radio, and television reporters seek to enlarge on the event. Such a situation occurred on August 8, 1986, the day after a male white shark, 5.1 m, 1,556 kg (16 ft 9 in, 3,427 lb), was caught on hook and line off Montauk, New York. Indeed, this is the largest white shark reported for our Atlantic coast, and the largest caught on rod and reel. But the specimen hardly compares with the 19 ft 6 in shark caught off Western Australia. We welcome future reports of large white sharks, but respectfully request the press to refrain from calling until at least a 19-footer becomes available! Since almost all reliable reports of monster whites come from Australia, we suspect we may have a long wait before getting such a call.

The white shark is an efficient and versatile swimmer capable of cruising long distances at low speed and making attacks at high speed, occasionally jumping clear of the water. One attempt to track the shark with sonic tags indicated it traveled 190 km in 2.5 days at an estimated average speed of 3.2 kph (a little more than 2 mph).

The white shark dines on an array of marine life and scavenges carrion and garbage. It devours, often whole, small and large bony fishes such as menhaden and tuna, as well as other sharks, including houndsharks, requiem sharks, hammerheads, and spiny dogfish. Young white sharks, less than 3 m, feed primarily on fish and have more slender teeth than the larger adults, which appear to prefer marine mammals. Cetaceans, including harbor porpoises, dolphins, and floating whale carcasses, are eaten by the white shark, but pinnipeds, such as the northern elephant seal, harbor seal, California sea lion, and Steller's sea lion, are favored food items. Dr. John McCosker, director of the Steinhart Aquarium in San Francisco, has written about an adult white shark, weighing over 1,350 kg, harpooned in California in the late 1970s, that had two adult harbor seals in its stomach. Rodney Steel, author of *Sharks of the World,* reported that in 1976 a commercial fisherman in California caught a white shark 5 m long with the bodies of two whole sea lions, weighing 56 kg and 79 kg, in its stomach. Although basking shark flesh has been found among the stomach contents of the white shark, it is not known if the white shark will attack that species. It is believed that the flesh was carrion from a harpooned shark. Some sea birds, such as gulls and gannets, have been found among the stomach contents of white sharks, but they are not believed to be common fare, and it is unknown whether they were alive when taken.

Experiments to calculate the amount of energy the white shark derives from its food appear to indicate that it is highly efficient and may be able to stock up enough reserves when well fed to go without a meal for more than three months.

How does the white shark attack its prey? Generally by surprise, and from behind and below, sometimes in an accelerated headlong rush at 17 kph to 25 kph, sometimes inverted on its back or rolling to its side, according to the size and posture of the prey. Telltale scars on the underside of the midbody, haunches, and flippers of northern elephant seals, California sea lions, and other pinnipeds have been identified as those inflicted by white sharks.

What happens at first bite? High-speed photography of white sharks near Dangerous Reef, Australia, reveal five steps in a single surface attack (figure 48). The sequence occurs in less than a second. First, the shark raises its snout; second, it drops its lower jaw; third, the upper jaw protrudes, fully exposing the teeth; fourth, the lower jaw moves upward and forward, closing the mouth and causing the teeth to pierce the flesh of the prey; and fifth, the snout drops as the head returns to its normal position. At the moment the shark bites, it does not see its prey because its eyes roll backward within the sockets, presumably to reduce the risk of injury from the teeth and nails of the struggling pinnipeds. It is thought the shark depends in large measure on the electromagnetic sensors (ampullae of Lorenzini) in its snout to find its prey in the final moments of attack.

For all its reputation as a ferocious killer, maneater, and perpetrator of unprovoked attacks upon boats, does the white shark really seek out human flesh for food? The answer to this question is not clear. Most students of the subject would respond that man is not normal prey for white sharks. The two seldom occur in the same habitat, and when they do, attacks are not inevitable. Even when attacks occur, the victim is often released. The white shark will bite, perhaps on an arm or

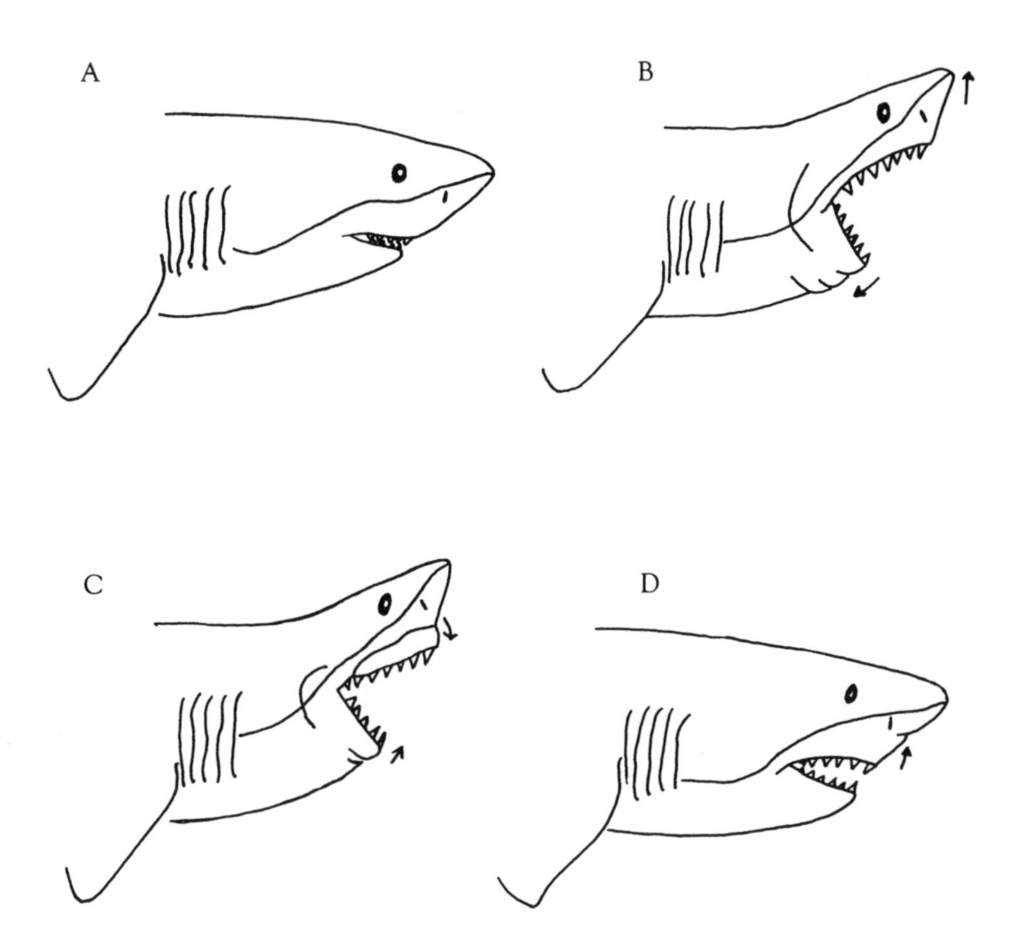

Figure 48. Sequence of jaw action in a surface attack by the white shark, Carcharodon carcharias: *a) just prior to biting; b) raising of head and simultaneous dropping of lower jaw; c) protruding of upper jaw and movement of the lower jaw upward and forward; d) lowering of head with return of upper jaw to its normal position. (From Tricas, 1985)*

leg, then move away and wait for its prey (man or beast) to bleed to death. Many whites will spit out human flesh after the initial attack.

Some scientists suggest that the "bite-and-spit" kind of attack may be the way this species samples potential food items, because it reduces the potential for injury to the shark by struggling and wounded prey. A powerful elephant seal could inflict much damage with its claws on the heavy-bodied, less maneuverable white shark. If left to bleed, the seal would weaken and no longer be able to fully defend itself when the shark returned for a second attack. Other scientists contend it is a case of mistaken identity that motivates attacks on humans. When viewed from below against a bright sky, skin divers in black neoprene wetsuits and swimmers on their short, manueverable surfboards appear similar in form to the dark silhouette of harbor or elephant seals, favorite foods of the white shark. Dr. McCosker made a study of the white shark's responses to the presence of dummies dressed to resemble divers in wet suits. Some of the dummies were placed upright on the ocean floor and others were attached to surfboards floating at the surface of the water with their legs and arms hanging off the sides. The sharks had to be tempted by bait to attack the dummies on the bottom, but no added attraction was necessary for the shark to attack the dummies on the surfboards. Indeed, the attacks were immediate and ferocious.

Although it is doubtful that white sharks deliberately leave their habitat to enter one where swimmers are normally found, the rogue theory of shark attacks suggests the possibility that once such a shark has savored human flesh, it is more inclined to seek out additional human victims. Accordingly and fortunately, human flesh rarely has been found among the stomach contents of a white shark.

More attacks on humans are attributed to the white shark than to any other shark species, even though most such reports come from tropical waters where the white shark is rare. It seems probable that the identification of the shark in many of these attacks is suspect. Quite possibly the culprits are the equally dangerous bull and tiger sharks, which are found most commonly in the tropics.

Is the white shark a boat biter? Indeed it is, and particularly those boats from which fish are being caught. In a few instances the attacks were so persistent that the boats sank. Such behavior may have something to do with the shark's electromagnetic sensitivity. Different types of metals that make up the hull, propeller, or rudder of a boat produce different electric fields in sea water, perhaps resembling the electric fields produced by a potential prey. The electromagnetic cues generated by the corroding portions of the boat thus may mislead the white shark. Divers inside the protective metal cages with bait tied to the bars have seen white sharks, apparently attracted to and approaching the bait, turn aside at the last moment of attack and bite the cage.

How many times have white sharks attacked humans? Probably not very many, but worldwide figures are unavailable. For the United States, off the coasts of California and Oregon from 1926 through 1984, records show only 59 shark attacks. Forty-six of these are attributed unequivocally to white sharks. Recordkeeping was probably not as complete or accurate for the early years covered by these data as it has been in more recent years. Of the 59 attacks, 12, including 2 fatalities, occurred in the period 1980 to 1984. In 10 of the 12 attacks, a white shark was implicated, and it is possible that one individual shark may have been responsible for more than one of the attacks. Along the coasts of New York and New Jersey, only one shark attack has been recorded in the 20-year period 1964–84, and the species involved is unknown.

At the risk of seeming flippant, we believe that white sharks are probably more in danger of being attacked by humans than we are in danger of being attacked by them. Sport and commercial fisherman kill 10 to 20 white sharks yearly off the California coast. If, as presumed, the white shark is not particularly abundant, requires many years to reach maturity, and gives birth to relatively few young, it may be possible in a relatively short time for man to so deplete their numbers by heavy fishing as to cause their extinction. Would it not be ironic, as is presently true of the maneating tigers that now face extinction, if man one day should find himself protecting the white shark? There is no doubt that white sharks can be aggressive and dangerous, and that surfers, divers, and swimmers should be wary in coastal waters where these sharks are known to occur. But to characterize the white shark as a totally sinister, rampaging monster, dripping with blood, is unwarranted.

Further readings

Ainley, D. G., et al., 1985; Cailliet, G. M., et al., 1985; Carey, F. G., et al., 1982; Casey, J. G., et al., (5) 1984, (8) 1987; Casey, J. G., and H. L. Pratt, Jr., 1985; Klimley, A. P., 1980; Lea, R. N., and D. J. Miller, 1985; McCosker, J. E., 1981, 1985; Randall, J. E., 1987; Taylor, L. R., 1985; Tricas, T. C., 1985; Tricas, T. C., and J. E. McCosker, 1984.

Whale Shark

In contrast to the ferocious white shark is the docile, colossal whale shark, *Rhincodon typus* (frontispiece), the largest of the living fishes. For many years the whale shark was thought to be the only member of its family, the Rhincodontidae, but recent taxonomic studies have enlarged the family to include the zebra and nurse sharks. It seems curious that this enormous beast is a member of the same family as the much smaller nurse shark, but the anatomical characteristics they hold in common are sufficient to bind them in a close relationship. The gentle nature of the whale shark, which is even less aggressive than that of the other members of its order (Orectolobiformes), serves perhaps, as behavioral evidence of a common bond among the species of the group.

Reports of the whale shark's maximum size vary because so few have actually been caught, weighed, or measured. Estimates based on records of sightings indicate that these sharks may grow to be 18 m (59 ft) or more, though the most recent specimen accurately measured was 12.1 m (39.9 ft). Ichthyologist and diver Dr. Eugenie Clark, author of popular accounts about her studies on sharks, attached herself to and "rode" a huge whale shark she estimated to be 12.5 m in length.

Whale sharks occur worldwide, but are restricted to tropical and warm-temperate seas, and are notably absent from the Mediterranean. They sometimes swim close inshore, but more often are seen in the open ocean, either singly or in schools consisting of hundreds of individuals. They are grayish or brownish above, whitish below, and strikingly marked with pale crosshatched stripes and a proliferation of pale spots, which in adults may be as large as the palm of the human hand.

Whale sharks are one of the few shark species that are filter feeders, dining on plankton, tiny shrimps, and small fish. Equipped with thousands of tiny teeth arranged in over three hundred rows in its massive jaws, this shark neither chews nor bites its food. Instead, it inhales water and food into its mouth, straining and draining the water out through its gills. It feeds while cruising slowly near the surface of the water or, alternatively, sometimes orients its body vertically, with its head above the surface, then lowers it so that huge volumes of water with both small and sometimes even large prey are sucked into its mouth. These sharks have been seen to evert their stomachs through their mouths. It is speculated that they

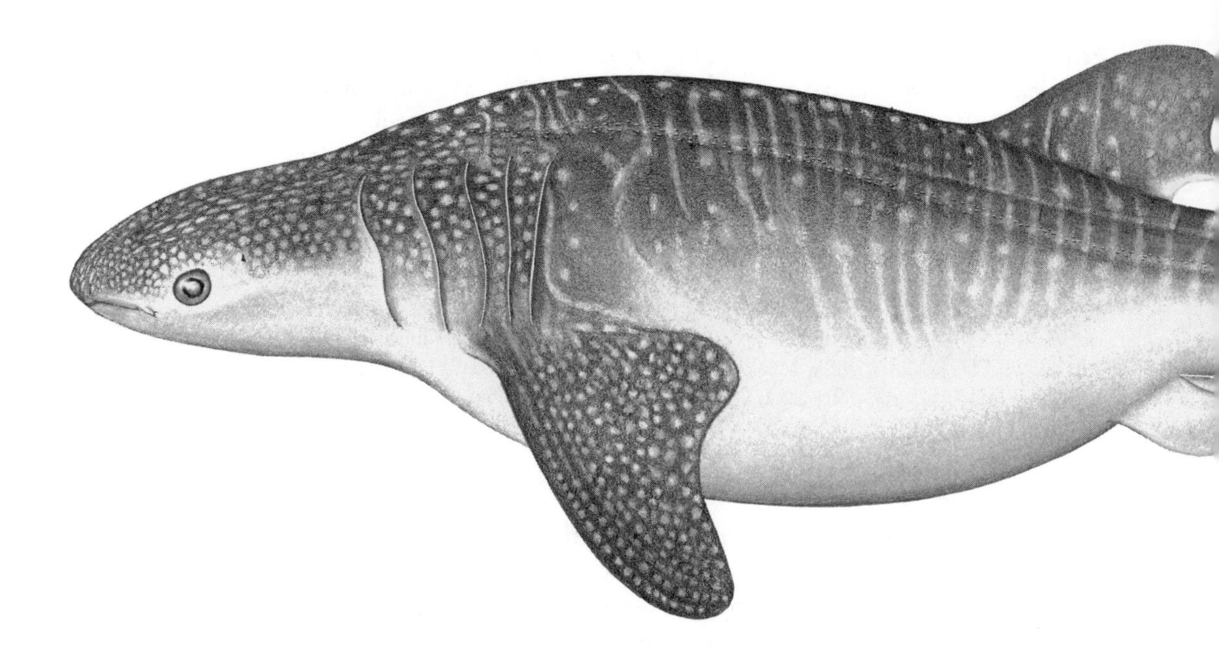

either cannot digest some of the food taken in, such as turtle shells, or that they are showing stress.

It was once assumed that whale sharks were egg layers (oviparous) but some evidence challenges this view. In 1953, a large egg case of a whale shark containing a live embryo (figure 49) was found in the Gulf of Mexico. This egg case was 30 cm long, 14 cm wide, and 9 cm thick. The embryo had a wide mouth and stripes and spots like those of the adult, but it retained part of its yolk sac, indicating it was not ready to be born. The egg case was extremely thin and had no tendrils, as are commonly found on the egg cases of oviparous sharks such as the bullheads. Also, the presence of umbilical scars on larger individuals has suggested that the whale sharks might bear their young alive (ovoviviparous), a mode of reproduction similar to that found in nurse sharks, in which the egg cases are kept in the uterus until the embryo is mature enough to hatch. One explanation for the free egg case is that it may have been aborted rather than laid. As many as 16 egg cases have been found in the uterus of an adult female whale shark, but much more information is needed to describe with certainty exactly how this species reproduces.

Further readings
Baughman, J. L., 1955; Garrick, J. A. F., 1964; Gudger, E. W., 1915, 1941; Harding, J., 1969; Wolfson, F. H., 1986.

Figure 49. Near-term embryo of the whale shark (actual size: 350 mm total length) removed from its egg case. (From Garrick, 1964)

Hammerheads

Truly weird and wonderful! By virtue of the unique conformation of their head, large size (over 2 m for three species), and because some are dangerous, the hammerheads (plates 8c and 9a) must be included among the supersharks. Unmistakable, their head in outline is boomerang-shaped, spade-shaped, or, more commonly, shaped like a double-bladed axe. The hammerheads comprise their own family, Sphyrnidae, and each of the eight species can be identified on the basis of head shape alone, although the differences in head shape of the larger species are subtle and may confuse the layman.

What is the purpose of this strangely shaped head with eyes and nostrils at either end? No one is certain. One theory proposes that the head acts as a bowplane to increase the sharks' maneuverability and give them added lift when swimming forward. Another theory emphasizes the possible sensory advantages of the arrangement of the wide, flattened head with widely spaced eyes and greatly enlarged nostrils. The increased surface of the head provides more room for greater numbers of lateral-line canals and ampullae of Lorenzini, enhancing the hammerheads' electromagnetic senses and response to water displacements and pressure. Located at the tips of the head, the eyes are thought to scan the scenery more precisely and with a greater range than if they were located behind the nostrils as in most sharks. The expanded nasal area probably serves to increase olfactory sensitivity. Recently, a great hammerhead was videotaped sweeping its head back and forth over the ocean bottom like a giant metal detector searching for a stingray buried in the sand, which it located and ate after circling for a few minutes.

The family Sphyrnidae is found around the world in shallow coastal waters, but there are interesting gaps in the distributions of the species, particularly the three largest species, which have the broadest distributions. The smooth hammerhead, *Sphyrna zygaena,* has an essentially north-south disjunct distribution: it occurs in the subtropical and temperate seas of continental coasts on either side of the equator, but is almost unknown in equatorial latitudes. In the great expanse of the central Pacific, it is known only in the Hawaiian Islands. In contrast to the smooth hammerhead, the scalloped hammerhead, *Sphyrna lewini,* is common in the equatorial latitudes along continental coasts, but similar to *S. zygaena,* is known only in the Hawaiian and Society islands in the central Pacific. The third large species, the great hammerhead, *Sphyrna mokarran* (plate 9a), also occurs along most warm continental coasts but has not been reported from the Indonesian-Philippine area or the central Pacific, except for the Society Islands. All the other species are small, and with the exception of the winghead shark, *Eusphyra blochii,* are variously restricted to the coastal waters of the Atlantic and eastern Pacific. *Eusphyra blochii* is limited in distribution to the northern coasts of the Indian Ocean, including the Red Sea and Persian Gulf, and to the coasts of Indonesia and Malaya eastward to the Philippines and south to northwestern Australia.

Hammerheads feed on bony fishes, other sharks, crustaceans occasionally, and in particular, on stingrays. It is not unusual to find individuals with venomous stingray barbs protruding from their mouth, throat, and tongue, evidence of a recent stingray dinner. The mouth and jaw of one specimen contained 96 barbs. Obviously, the barbs seem not to bother the sharks, but little is known about their effect and how, or if, the shark rids itself of them.

Some species travel in large schools and migrate, moving poleward in the summer. Hammerheads are viviparous, with the great hammerhead having as many as 42 young and the bonnethead as few as 4 to a litter (plate 15: a and b). Litter size seems to depend on the size of the mother and the maximum size attained by the species.

The smallest of the hammerheads is the bonnethead, *Sphyrna tiburo* (plate 8c), which attains a length of about 1.5 m and is harmless. The largest species is the great hammerhead, *Sphyrna mokarran,* which reaches a maximum length of about 6.1 m. Most adults of this species, however, range from 3 m to 5 m (which is large enough!). Several attacks on both swimmers and boats have been attributed to the great hammerhead, but it is not certain if they were made by this shark or its two large relatives, the scalloped hammerhead and smooth hammerhead. Even though potentially dangerous, hammerheads are not particularly aggressive and usually retire in the presence of divers wearing scuba. Although divers without gear have approached hammerheads without harm, we recommend caution when interacting with these beasts.

Groups of individuals of some species of hammerhead sharks show behavior patterns that belie the so-called unpredictability of sharks. In particular, studies by ichthyologists Drs. A. Myrberg and S. Gruber of a colony of 10 bonnetheads at the Miami Seaquarium, and observations by Drs. A. Peter Klimley and Donald R. Nelson of schools of scalloped hammerheads in the Gulf of California, indicated the formation of special relationships among the members within a group and revealed that certain types of behavior were repeated often enough to be anticipated in given situations.

The bonnetheads were observed in a seminatural pool over a 6-month period during which 18 postures and patterns of movement were noted and given names. Of these 18, about half were repeated often and seemed to have social significance. Myrberg and Gruber described the different modes of swimming as "patrolling" (swimming in a straight line), "maneuvering" (making sharp lateral turns), and "explosive gliding" (a series of rapid tail beats followed by a long glide). Other behaviors by single individuals included "head-shakes," "jaw-snaps," "chafe" (rolling body against the bottom), "gill-puffs," "clasper-flexion," "clasper-flexion thrust," "circling-head-to-tail," and "approach-over-the body."

An important finding of the study was the relative lack of aggressiveness among the members of the group, which seems to indicate that these sharks live peacefully together in nature. Although no fighting occurred, there were several cases ("hits") of a resident hammerhead's contacting a newcomer during the first hour after the newcomer was introduced into the pool. Another behavior called the "hunch" also occurred at least five times in the pools. During a hunch a resident shark arched its back, lowered its tail, and raised its head as it passed the newcomer. Other actions with social relevance included: "circling head to tail" (two sharks moving in a tight circle); "turn back" (a shark swimming in one direction reversed itself abruptly to follow another shark); "follow" (one shark closely following another); "follow-formation" (a procession of three or more sharks following a leader); and "give way" (yielding the right of way to avoid a head-on confrontation). From this study, a subtle social organization based on size was found to exist. Even though sex did not determine position in the group, the large males were deferred to by the other sharks.

Despite the potential danger of swimming with a school of scalloped hammerheads, scientists have risked this in the Gulf of California over a period of several years in order to observe the behavior of these sharks. Scalloped hammerheads are large sharks, reaching 3 m in length, and they are possibly the most abundant of the wide-ranging species of hammerheads, forming large schools with as many as 100 individuals (the bonnethead shark, a much smaller species, is probably the most numerous of the hammerheads, reportedly occurring in schools of thousands). Various other large species, such as the bullhead, gray reef, sandbar, basking, and whale shark, are known to gather in groups of three or more individuals, and some researchers think that grouping behavior occurs somewhere in the life cycle of most sharks.

Scalloped hammerheads may form true schools: they swim in a polarized way (all moving in the same direction), maintaining a constant distance from one another and changing direction at the same time. In the Gulf of California study, stereographic photographs taken of these groups seem to indicate that the larger sharks stay along the edges of the group, and that most of the schools were composed predominantly of females, which outnumbered the males by as much as 3.8 to 1. From time to time, single sharks in the school engage in various kinds of acrobatic behavior, tilting their bodies laterally, accelerating upward or downward, and shaking their heads from side to side or just to one side. One of the more spectacular displays is "corkscrewing," a rapid 360 degree twisting of the body while accelerating into a circular path, sometimes hitting another shark in the process. Other behaviors include jaw opening and clasper flexing. It is possible that aggression and courtship may be involved in these displays.

Why do scalloped hammerheads group? There are at least four theories, but none of them appears to be viable: courtship and mating, protection against predators, increased efficiency in swimming against strong currents (has to do with the physics of fluid dynamics), or increased efficiency in finding and capturing prey. Most hammerhead schools consist of several small, immature sharks, so reproduction is not necessarily a factor. Large predators, like the tiger or white sharks, have not been seen during the day in the same habitat as schooling hammerheads, so defense may not be a factor. Furthermore, scalloped hammerheads swim together when currents are absent, so there may be no hydrodynamic advantage involved. It may be advantageous for sharks to gather near food sources and cooperate in locating prey, but it has not been determined that the scalloped hammerhead forms groups for this reason. No scalloped hammerheads have been seen to feed during the day; and so feeding may take place only at night, when most fish schools tend to lose their cohesiveness.

What the sharks do at night may provide insight into the larger question of why they group during the day. There is also the possibility that the hammerheads do not go about their usual business during the day when ichthyologists are swimming nearby. Or just maybe, schooling is on occasion simply a gregarious reaction to, or recognition of, one's own species.

Further readings
Bass, A. J., et al., 1975a; Gilbert, C. R., 1967; Klimley, A. P., 1981; Klimley, A. P., and D. R. Nelson, 1981; Myrberg, A. A., Jr., and S. H. Gruber, 1974.

Bull Shark

Of all the sharks swimming the continental coasts of tropical and subtropical seas, the bull shark, *Carcharhinus leucas* (figure 26), may be the most dangerous. In number of attacks on humans, it is one of the three species of sharks most often implicated. It may not have the white shark's reputation, but its large, heavy body, huge jaws, and very large teeth make it meter for meter just as formidable. Even though it appears to move slowly when cruising the shallows inshore, it is capable of fast, agile movements when it wants to attack prey. In Natal, South Africa, the bull shark is particularly notorious as an aggressive species, and it is caught four times more frequently than either the white or tiger sharks in the protective antishark nets set off the beaches.

The species is relatively large, purportedly growing to a length of 3.4 m, but with an actual record of up to only 3.2 m (based on a report from Brazil). Individuals over 3 m are rare.

The bull shark is the only shark known to live in the saltwater of the oceans, the brackish water of estuaries and river mouths, and the far upper reaches of freshwater rivers and lakes. Even its newborn young and juveniles enter freshwater without difficulty. It travels from the Caribbean 100 km up the Rio San Juan to Lake Nicaragua, where it was long thought to represent a different land-locked, endemic species. It has also been reported from Lake Ysabel in Guatemala, the Mississippi and Atchafalya rivers in the southern United States, 3,700 km up the Amazon River to Peru, the Gambia River on the middle upper west African coast, more than 550 km from the sea up the Zambezi River on the middle lower east African coast, the Tigris River of Iraq, the Hooghly River of northeastern India, and Lake Jamoer in New Guinea. These are but a few of the bodies of freshwater frequented by bull sharks. As recently as August 12, 1985, a large, 2.7 m bull shark weighing about 216 kg was caught in the Chesapeake Bay along the western North Atlantic seaboard.

The bull shark's tolerance of a wide range of salinities in warm continental waters does not, however, enable it to maintain populations worldwide. Among the many islands of the whole central Pacific, it has been reported only once, from Rangiroa, in the Tuamotus. It seems probable to us that the bull shark is primarily restricted to warm continental coasts and associated inland waters. Quite possibly it has a brackish or freshwater ecological requirement, as indicated by the frequent presence of near-term females and juveniles in such waters.

Most of the information on the life history of the bull shark is sketchy, often contradictory, and known mainly from studies on the populations in Lake Nicaragua and adjacent coastal areas and along the coasts of Florida and South Africa. In Florida, mating appears to take place in June and July, whereas along the coast of Nicaragua it probably occurs all year. A gestation period of 10 to 11 months has been reasonably well established for Florida and appears to be true also for Nicaragua. The species is viviparous; only the right ovary produces eggs; and females bear from 1 to 13 pups, but usually about 5 or 6.

Bull sharks grow slowly and mature late. Lake Nicaraguan bull sharks grow rapidly during their first two years, 16 cm to 18 cm per year, but subsequently slow to 9 cm to 10 cm per year. Studies in the northern Gulf of Mexico indicate that the period of fastest growth (15 cm to 20 cm/yr) extends over the first 5 years, with the rate decreasing gradually to about 4 cm to 5 cm by the time they are 16 years old. Off South Africa, sexual maturity is reported variously to be reached at the age of 6 or 15, while in the northern Gulf of Mexico it is not reached until at least age 14 for males and about 18 for females.

How long bull sharks live in the wild is unknown. It has been estimated, however, that female bull sharks in Lake Nicaragua have a life span of 16 years; males only 12 years. In the northern Gulf of Mexico, based on a study of ring counts in the vertebrae, a male, 2.45 m long was estimated to be 21.3 years old, and a female, 2.68 m long, to be 24.2 years old. Captive bull sharks in the Durban Aquarium in South Africa have survived 15 years.

Much information on the life history of sharks is obtained from tagging studies, which in the case of bull sharks have been unsuccessful. During the period 1963 to 1986, the National Marine Fisheries Service Cooperative Shark Tagging Program on the East Coast succeeded in tagging only 283 bull sharks, of which only three were recaptured, hardly enough to provide useful information.

Bull sharks feed on a variety of prey and are almost as omnivorous as the tiger shark, the ocean's junk food eater; however, the bull is less likely to eat indigestible garbage. It favors bony fishes and other sharks, particularly young ones in the nursery grounds. In the sea, it will also kill and eat dolphins and sea turtles. In rivers, it has been known to kill hippopotamuses. Even the remains of large land animals, including antelope, cattle, dogs, and humans have been found among its stomach contents.

Dr. Leonard Compagno, who was ichthyological consultant to the film *Jaws*, believes that the bull shark is quite possibly the most dangerous of all sharks, and that many attacks attributed to other species may actually have been committed by bull sharks. The famous, closely spaced series of five attacks—four fatal—that occurred in New Jersey in 1916, have been attributed to the white shark. At least three of the attacks, however, occurred two miles up a narrow tidal creek, a habitat type in which no white shark has ever been confirmed to occur. A tidal creek is a likely place for bull sharks to be found, and few other species of sharks have been reported from such a habitat. No other western Atlantic shark species is known to enter tidal creeks.

Although bull sharks are dangerous to man, they have some redeeming features. Their flesh is used for human consumption; their fins provide the base for an expensive delicacy, shark-fin soup; their liver contains oil rich in vitamin A; and their hide is a source of leather. The species is a popular game fish in some areas and is prized for its strength and fighting ability.

Further readings
Bass, A. J., 1977; Branstetter, S., and R. Stiles, 1987; Jensen, N. H., 1976; Merrick, J. R., and G. E. Schmida, 1984; Myers, G. S., 1952; Snelson, F. F., Jr., et al., 1984; Thorson, T. B., 1971, 1972, 1976; Thorson, T. B., and E. J. Lacey, Jr., 1982; Thorson, T. B., et al., 1966.

Blue Shark

A veteran traveler, the blue shark (plate 5b) is known for its extensive wanderings in temperate and tropical seas. One blue shark tagged off New York was recaptured 16 months later, 5,984 km (3,740 miles) away, off the coast of Brazil, a long-distance record for the National Marine Fisheries Service Cooperative Shark Tagging Program and first evidence that blue sharks from the western Atlantic cross the equator. The previous distance record was for a blue shark recaptured after six years that had traveled across the Atlantic from New York to Liberia, 5,828 km away. The fastest rate of travel for a blue shark over a great distance is 2,764 km in 64 days, or an average of 43.2 km per day. The longest period between tagging and recapture of a blue shark is 8 1/2 years, for an individual tagged off North Carolina and recaptured near Georges Bank, east of Cape Cod, Massachusetts.

The blue shark, *Prionace glauca,* derives its common name from the brilliant blue color of its sides and back; its underbelly is bright white. A slender body, long pointed snout, long curved pectoral fins, and white-rimmed eyes that seem to stare impassively about its watery world, make this shark one of the more easily recognized species. A sinuous swimmer, it is the most abundant shark of the open ocean off the east and west coasts of the United States. It may be seen in large aggregations or alone (plate 12a), slowly gliding or basking near the surface with its pectoral fins outspread and dorsal fin and tip of tail above water. The blue shark is not so passive when pursuing prey and is capable of swift and powerful movements, even jumping out of the water when hooked.

Studies of blue shark swimming behavior in the wild, using acoustic telemetry (the tracking of sound-producing tags) show that the sharks not only swim at the surface, but consistently dive from the surface to depths as great as 500 m. The dives occur at two-hour intervals and last about an hour, with the deepest dives occurring during the day and the shallowest at night. The sharks actively swim downward during most of a dive, with brief interspersed periods of gliding. It is believed that this swimming pattern represents a strategy for maximizing the encountering and capture of prey.

Blue sharks are viviparous. The young are born after a gestation period of 9 to 12 months, the number of pups varying widely, from as few as 4 to as many as 135 per litter. The skin of the female blue is more than twice as thick as that of the male, thicker than the male's teeth are long, perhaps to protect her from the male's toothy advances. Even though courting and mating blues have not been observed, toothcuts on adult and subadult females, and lack of wounds on males, indicate a mating ritual that involves biting. One can readily identify the sex of a blue shark solely from the presence of such wounds.

Blue sharks mature at a relatively early age. In the western North Atlantic, males mature at about 4 to 5 years, females at 5 to 6 years. Females have a subadult stage, from age 4 to 5, when they begin sexual activity and copulate, usually in the late spring to early winter. They store the sperm until the following spring when the sharks are 6 years old. Then the ovary (only the right one is present and functional) and oviducts enlarge and become differentiated. Fertilization then occurs and the pups are born in the mother's seventh year.

The largest blue shark reliably reported was about 3.8 m long. Age and growth studies of the blue shark from the eastern Pacific estimate an individual 2.7 m long would be about 20 years old, so a shark 3.8 m long would be much older. How long the blue shark lives is unknown, but based on the sizes most often caught, few probably live more than 20 years.

The blue shark feeds on small bony fishes and devours large numbers of squid (plate 11a). In the western north Atlantic the most important food items inshore are bluefish and hake, whereas offshore, squid is the major food item. Feeding rates of the blue shark were unknown until recent sonic tracking experiments showed that under natural conditions blue sharks can digest about 1.2 kg of mackerel in 24 hours, or 3 percent of their body weight per day. To obtain this information, blue sharks were attracted to a boat by chumming, then fed a measured amount of food and tagged. The sharks were tracked until they surfaced hours later, at which time they were harpooned, brought aboard, and their stomachs examined to determine how much of the food had been digested.

Various strategies have been developed by the blue shark for feeding on huge masses of squid. Sometimes the shark slowly sweeps its head and tail back and forth, catching large amounts of prey in the corners of its mouth. At other times it may charge quickly through the dense schools with its mouth wide open. It also can assume a vertical position beneath the squid and accelerate upward to engulf them. Blue sharks are often seen in large feeding frenzies on whale carcasses.

Known to be dangerous, blue sharks will attack humans and even boats. Spearfishermen and divers should be particularly cautious in their presence, especially if speared fish are in the area. Sometimes the sharks just appear to be curious, hovering and circling divers for more than a quarter of an hour before departing. At other times, such curiosity results in an attack.

Further readings
Cailliet, G. M., et al., 1983; Carey, F. G., and J. Scharold, 1987; Casey, J. G., et al., (1) 1980, (7) 1986; Casey, J. G., and C. E. Stillwell, 1983; Pratt, H. L., Jr., 1979; Scharold, J., and F. G. Carey, 1985, 1986; Strasburg, D. W., 1958.

Shortfin Mako

A superlative game fish, the shortfin mako, *Isurus oxyrinchus* (plate 6b), embodies all the characteristics prized by sport fishermen: beauty, power, aggressiveness, grace, and athletic prowess. Made famous in Ernest Hemingway's *The Old Man and the Sea,* the shortfin mako is one of the most distinctive sharks: steely blue above and white below, with a long, pointed, conical snout, large dagger-like teeth visibly protruding from the lower jaw, horizontally flattened keel (reinforcing ridge) at the base of the tail, and dark menacing eyes—an incomparable spectacle for the fisherman lucky enough to catch one. When hooked, it has the strength to leap at least 6 m (20 ft) into the air (figure 21), which, based on theoretical computations, would require it to be traveling more than 35 km/hr (22 mph). It is believed to be the fastest of all the sharks, although no accurate measurements of its speed have been made. Certainly, it holds the speed record for long-distance travel: approximately 2,130 km in 37 days or an average of about 58 km (36 miles) per day.

The shortfin mako is known to attack boats, particularly if it is on an angler's line, sometimes jumping into a boat, snapping and thrashing about and creating such havoc that frightened anglers have been known to dive into the water for safety.

Found offshore in tropical and warm temperate waters of all the oceans, the mako feeds on a variety of fishes including not only the fast-moving tunas, swordfishes, and other sharks, but also herring, cod, and squid. Marine mammals, however, do not seem to be an important food item in the mako diet.

Like the great white, the mako maintains an elevated body temperature 7 to 10 degrees Celsius higher than the water temperature, which enables it to capture and feed on the swiftest fishes. In the northwestern Atlantic, it ranges from the Grand Banks to the tropics and is common from Cape Cod to Cape Hatteras. In these waters a major item of its food is bluefish (*Pomatomus saltatrix*), followed by squid. Examination of the stomach contents of 399 shortfin makos indicated that 77.5 percent of the diet was bluefish. Both males and females consumed the same kind and amount of food. Bluefish are widely distributed in the coastal waters on both sides of the Atlantic, but are absent in the Pacific Ocean, except for the coast of Australia, and in the Indian Ocean, except for the southwestern and southeastern sections. Because the shortfin mako occurs worldwide, it must concentrate its feeding on organisms other than bluefish in areas where the latter is absent.

Investigators working in the northwestern Atlantic and the eastern Pacific have given widely different estimates of the growth rate of the shortfin mako. In the Atlantic, the shortfin mako appears to grow unusually fast, almost twice as fast as in the Pacific (California), or as the porbeagle (another lamnid), and much faster than requiem sharks. The differences are the result of different scientists' interpretations of the number of rings (annuli) added yearly to the sharks' vertebral centra. In the Pacific study, one ring, and in the Atlantic study, two rings, were believed to form each year, although the Atlantic investigators considered the possibility that rings might not form annually in older individuals.

It is probable that shortfin makos grow at about the same rate in both the Atlantic and Pacific, and that only one, if either, of the two assumptions regarding number of rings formed each year is correct. Since the maximum reliably measured sizes (Atlantic males 2.8 m, females 3.28 m; Pacific males unreported, females 3.24 m) attained by shortfin makos in the Atlantic and Pacific are about the same, estimates of ages attained by the two studies should also vary by a factor of two. In the Atlantic, however, the longest shortfin mako males and females were estimated to be 4.5 and 11.5 years old, respectively, whereas in the Pacific the oldest shortfin mako was believed to attain an age of as much as 50 years.

In Australia, female shortfin makos become mature at 2.8 m and give birth to from 4 to 16 young that are approximately 60 cm to 70 cm long. Russian reports indicate that the embryos are attached to the oviduct by placentas, but the Australian study found that all the embryos in several females lacked placentas or placental scars and had eaten yolk. The indications are that the young feed on unfertilized eggs in the oviduct, just as occurs in other lamnids, including the longfin mako.

Although the shortfin mako is considered dangerous, it is commonly found only well offshore where there are few swimmers and divers. Reasonably verified, unprovoked attacks on swimmers resulting in wounds are rare. One such attack (nonfatal) occurred in 1974 in the Gulf of Elat (Red Sea), some 500 m offshore; swimmers there seldom venture out more than 100 m. On the other hand, there have been several reports of mako attacks on boats as verified by teeth left embedded in hulls.

Shortfin makos are not only popular for their fighting ability, but also for their outstanding qualities as a foodfish. Those in the know consider shortfin makos to be the choicest fare of all sharks. It is only recently, however, that mako has become common in restaurants along the coasts of the United States. Some cookbooks now include recipes for sauteed mako fillets and mako kabobs.

Further readings
Cailliet, G. M., et al., 1983; Gilmore, R. G., 1983; Pratt, H. L., Jr., and J. G. Casey, 1983; Randall, J. E., and M. F. Levy, 1976; Stevens, J. D., 1983; Stillwell, C. E., and N. E. Kohler, 1982.

Basking Shark

A mild-mannered mammoth of the shark world, the basking shark, *Cetorhinus maximus* (figure 50), is the second largest fish in the world, surpassed only by the whale shark. Most basking sharks are under 9 m (30 ft) in length, but some may grow to 15.2 m (50 ft). Many of the tales of sea serpents and monsters have their origin in sightings of lines of basking sharks cruising single file, snout to tail, near the surface of the water, or from the decomposing remains of huge basking shark carcasses found washed up on beaches. Because the basking shark's skull is relatively small compared to its body length, it seems unbelievable to the layman that these carcasses, sometimes dismembered, are those of a shark rather than some unknown beast. But when the fanciful stories are investigated, time and time again the animal is identified as a basking shark.

Its massive size, pointed snout, subterminal mouth with minute hook-like teeth, and enormous gill slits, which almost completely encircle the head and are filled with bristle-like gill rakers, distinguish the basking shark from other species. The sight of a gigantic shark the size of a bus can be terrifying. But its appearance belies its harmless nature.

The basking shark is a plankton feeder, slowly, powerfully swimming near the surface at a steady rate of about 2 knots, with its mouth open when feeding (plate 11b) and its gills distended as it filters as much as 1,800 metric tons (1 metric ton = 2,200 lb) of nutrient-rich water an hour. The water flows through the mouth to the pharnyx, then out through the gills. Inside the gill chamber on the gill arches, thousands of gill rakers, coated with a sticky, mucous secretion, sieve the plankton, mainly tiny crustaceans, from the water. From time to time the shark will close its mouth, constrict its gills, and swallow these organisms. The basking shark depends solely on the passive flow of water for ingesting its food, whereas megamouth and the whale shark, the only other shark filter feeders, may gulp or pump water through their mouths.

The basking shark inhabits the subpolar to temperate sea coasts of both the northern and southern hemispheres, but is absent, except for strays, from the warm waters in between (figure 51). It is occasionally seen cruising in the open

Figure 50. This basking shark, Cetorhinus maximus, *is a 5.7 m male caught off South Africa. (J. L. B. Smith Institute of Ichthyology)*

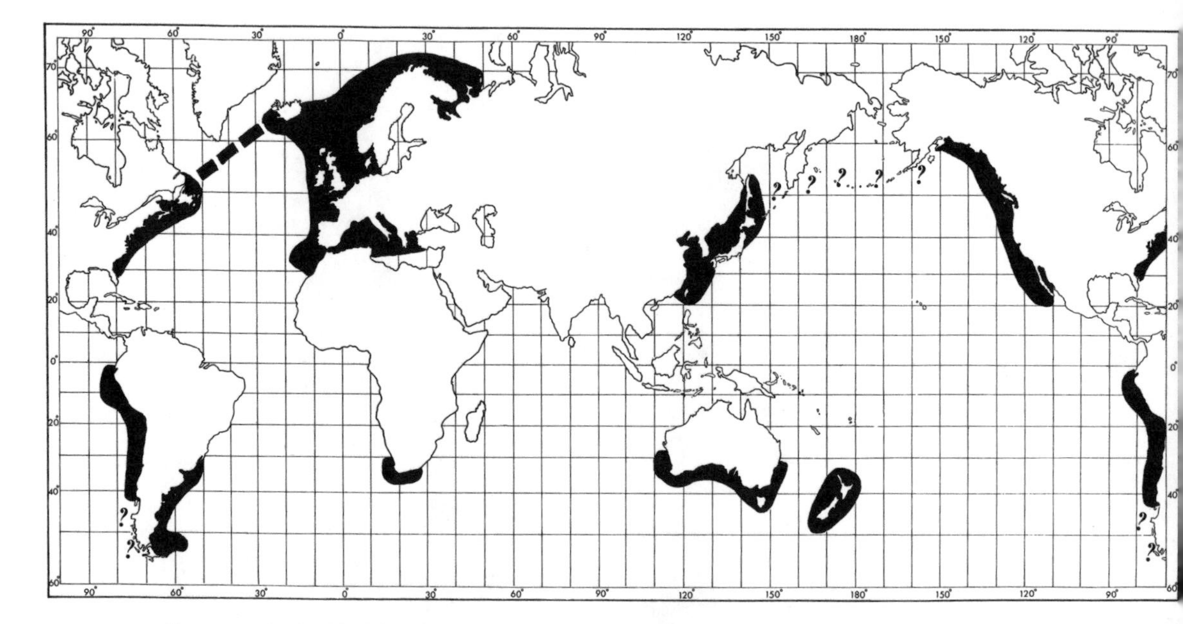

Figure 51. Generalized distribution of the basking shark, Cetorhinus maximus *(dark dashes indicate presumed migration route). The basking shark is a bi-temperately distributed species, which occasionally strays into warm waters, for instance, the eastern Gulf of Mexico. Bi-temperate distributions are exhibited by several shark species. (After Compagno, 1984)*

ocean far offshore, as well as close to land, singly, in lines of two or more, or in schools of over a hundred individuals. The basking shark takes its name from habitually basking at the surface of the water with its dorsal and tail fins exposed, or lolling about with its belly up. Some scientists believe that basking sharks may normally inhabit deeper water, coming to the surface when food is especially abundant, temperatures favorable, or to engage in courtship and mating. Enormous as it is, this shark is reported occasionally to leap clear out of the water, possibly to dislodge parasites such as copepods and lampreys, a spectacular feat for a shark that may weigh as much as 3.6 metric tons. Reports of jumping basking sharks, however, may be the result of misidentification. A shark of such size that can only accelerate to 4 knots when harpooned is unlikely to have enough power to breach like the great whales, and the animals in question might be dolphins or thresher sharks.

A highly migratory species, the basking shark appears at certain localities at certain seasons, then disappears, its whereabouts and behavior unknown. In the spring it is found off the eastern coast of North America from North Carolina to New York, then migrates in the summer to the northern part of its range, New England and Canada, and vanishes in the autumn and winter. A similar disappearing act takes place off the British Isles, although a few individuals, mostly males, may be found in that locale throughout the winter. Large numbers appear in the spring, and in the summer courting and mating probably occur, as evidenced by scars on captured animals. By autumn most of the population disappears again. In contrast, in the eastern Pacific, basking sharks are abundant off central California most of the year, but during June and July are rarely seen. Satellite tracking of long-range radio tags attached to these sharks could provide answers to where they go and what they do when they disappear.

Because the basking shark periodically sheds its gillrakers (during the colder months in the North Atlantic) and because the amount of plankton varies with the season, the question arises as to how the shark feeds during the winter. One theory suggests that the shark hibernates on the bottom until its gillrakers regenerate and the plankton blooms again. Another explanation is that the shark feeds on the bottom until the gillrakers grow back, and still another theory supports an overall reduced rate of activity with the shark's energy needs fueled by the oil stored in its huge liver. The liver may be as much as 25 percent of the shark's total weight and produce well over a thousand liters of oil.

Very little is known about reproduction in this shark. The species is probably ovoviviparous like other lamnids, and the embryos in the oviducts probably feed on unfertilized eggs produced by the mother. Adult nonpregnant females have been found with several million tiny eggs in their ovaries, but no pregnant females have been captured, nor has the birth process been observed. Because the smallest basking shark caught in the wild was reliably reported as 1.7 m long, and because of an unconfirmed report of a fetus of the same length, it is believed that a newborn basking shark is about that size at birth. If the figures are accurate, the size at birth is greater than that of any other shark. Young juvenile basking sharks differ in one feature from their parents: they have a pointed snout with a horn-like protuberance at its end, giving them a rather elephant-like appearance.

The basking shark is thought to grow slowly and mature late, males possibly at 12 to 16 years or more. It is believed to have a long gestation period, three and a half years, and to give birth to relatively few young. (If the estimated gestation period proves to be true, it would be 18 to 20 months longer than that of the spiny dogfish, *Squalus acanthias,* which has the longest-known gestation period of any vertebrate.) These factors, including the small size of existing populations, make the basking shark susceptible to overexploitation. The species is actively sought off the coasts of Japan, China, and Britain where significant fisheries exist. Its habit of slowly cruising near the surface makes it easy prey for harpooners.

The basking shark is valued for its meat, over a ton in an 8 m specimen, its fins, used in shark-fin soup, its hide, used for leather, and for its carcass, which is ground into fish meal and used for both human and livestock consumption. Its liver is particularly sought for its oil, which contains certain important components. These include the hydrocarbons pristane, used in lubricants and skin softeners (emollients), and squalene, used in cosmetic preparations such as skin-conditioning creams. Pristane is inert biologically, has a very low congealing temperature (−60 degrees Celsius), and is useful in releasing ice and frost on engines. One of the derivatives of squalene is the saturated hydrocarbon squalane, used in fixatives for perfumes, in aerosol hair-setting products, lipsticks, moisturizers, and lubricants for delicate instruments. The supply of basking sharks is highly limited and there is only one processor of basking shark-liver oil in the United States.

How dangerous is the basking shark? Not at all, at least in terms of what it eats. Only its mammoth size and unexpected movements pose a danger to divers and fishermen. A quick swipe from one of its fins is as dangerous as being hit by a bus. A brush with its thick, rough skin, armed with pointed denticles, can cause painful cuts and scrapes.

Further readings
Baldridge, H. D., 1972; Davis, C., 1983; Robeco Chemicals, Inc., 1984; Rosenthal, M. L., 1964; Springer, S., and P. W. Gilbert, 1976; Squire, J. L., Jr., 1967.

Thresher Shark

The thresher shark, *Alopias vulpinus* (plate 9b), is aptly named. It has an extraordinarily long scythe-like tail that may be as long as the rest of its body. What purpose could such a tail have? It is commonly reported that the thresher has a specialized method for capturing prey. It vigorously splashes the water with its powerful tail while circling schools of fish, herding them close together, then whipping its tail back and forth, stunning and killing its prey. Threshers are often found hooked on longlines by the upper lobe of the tail rather than through the mouth, perhaps while trying to stun bait. Longlines are a special type of fishing gear, consisting of a line up to several miles long with other lines attached and spaced along the "long line."

Alopias vulpinus is one of only three species in the family Alopiidae; the other two species are the bigeye thresher (plate 9c) and the pelagic thresher. An energetic, stubborn fighter, sometimes leaping out of the water, *Alopias vulpinus* is classified as a game fish by the International Game Fish Association. The colors of its back and sides are dark brown to blue gray to black with a metallic luster, shading off to white on its belly.

The thresher occurs worldwide, usually far offshore in temperate to tropical oceans and at depths from the surface down to 366 m, but where there are schools of bluefish, mackerel, needlefish, and squid, it may come close to shore in pursuit. Most of the young stay along the coast and can be found in shallow bays. In the northwestern Indian Ocean between January and May, for reasons unknown, males and females are predominantly segregated, occupying different depths and areas.

All the species in the Alopiidae are ovoviviparous and exhibit intrauterine cannibalism, the embryos feeding on each other and the unfertilized eggs (plate 14: b and c). The length of the gestation period of the thresher is unknown, although one report estimates a period of 9 months for the species in California waters. There are usually two to four pups in a litter, with the newborn measuring 1.1 m to 1.5 m. This shark appears to have a slow growth rate and a maximum total length of about 6 m (counting the tail, which may be longer than the body), although unverified reports give lengths up to 7.6 m. As body length increases, the relative size of the eyes and mouth decrease. The larger individuals are reported from the northeastern Pacific, and weigh as much as 454 kg. *Alopias vulpinus* appears to have a long life span, but its maximum age is unknown. It is has been estimated from growth curves that at a hypothetical total length of 6.5 m (larger than any verified report), a thresher may be as much as 50 years old.

Further readings
Bass, A. J., et al., 1975b; Bedford, D. W., 1987; Cailliet, G. M., et al., 1983; Gubanov, Y. P., 1972.

Tiger Shark

Like its jungle namesake, the tiger shark, *Galeocerdo cuvier* (plate 5c), a member of the family Carcharhinidae, is a large, fierce predator. It is easily recognized by its big broad head, blunt snout with prominent flaps on the nostrils, strongly curved serrated teeth with a deep notch, and very long pointed tail with strong reinforcing ridges (keels) at its base. It is the color pattern of the young tiger sharks, however, that has prompted its common name. The back of a young tiger shark is mottled with dark gray to black spots, which fuse into vertical stripes as the shark gets older and larger. These stripes tend to fade and may be absent altogether in the largest individuals.

The tiger shark prowls the tropical and warm temperate oceans throughout the world, swimming close to shore, sometimes in bays and near wharves, but also frequenting the open ocean, devouring all kinds of prey. In the Caribbean area, in particular, this species is one of the most abundant of the large sharks.

Of all the sharks, the tiger shark is the most unselective and opportunistic in the types of food it eats. It is also considered the most dangerous shark in tropical waters, with a reputation as a maneater. It consumes the most unusual items of any size, form, and material. Not only does it eat all kinds of bony fishes, other sharks, marine mammals such as seals and dolphins, and invertebrates such as lobsters, crabs, and conchs, but it scavenges animal cadavers—dead rats, cattle, dogs—and inedible objects such as tin cans (even an unopened can of salmon), metal, burlap and plastic bags. It attacks anything that floats or is thrown or falls into the sea. With its heavily calcified jaws and serrated teeth, the tiger shark has no problem biting through the heavy carapaces of sea turtles. It is one of the few sharks that prey on marine turtles and sea snakes. Even horses' heads and human limbs have been found in its stomach. All of these oddball food items are evidence that the tiger shark most commonly frequents shallow inshore waters, where its most common prey, however, are bottom fishes and invertebrates.

The tiger shark seems to be a night stalker, moving inshore into shallow areas to feed at night and retreating to deeper water during the day. When tiger sharks are seen near the surface during the day, they are the younger and smaller individuals. The species appears to be a loner even though it will join other tiger sharks to feed. It is slow and lethargic when cruising, but can quickly accelerate when attacking, and can be particularly tenacious and persistent in holding and fighting prey. Very few tiger sharks have been tagged and recaptured, and the information from tag returns is inadequate for discerning any pattern to their movements (plate 13c).

The following examples illustrate the apparently random travels of these sharks. Of 1,321 tiger sharks tagged from 1980 to 1987 by the National Marine Fisheries Service, only 56 have been recaptured. One shark tagged off Alabama was recaptured 65 months later off Cuba, approximately 1,000 km away, while a shark tagged off New York was recaptured six months later off Costa Rica, almost 3,000 km away. Three sharks tagged off New Jersey and Virginia were recaptured from 2 to 31 months later off North Carolina (by the same fisherman at the same place!), 260 km to 630 km away.

There are few detailed studies of the biology of the tiger shark. The most recent study indicates that mating occurs in the spring with a gestation period of about a year or slightly longer. Unlike most other carcharhinids, which are viviparous, the tiger shark is ovoviviparous, giving birth to litters of from 10 to 82 pups, usually more than 70 cm (up to 76 cm) long. The pups double in length the first year, after which they grow more than 20 cm a year until they are near maturity.

Growth and maturity vary in different populations. In the Gulf of Mexico, tiger sharks grow about 4 cm a year faster during their first 4 years of life than they do in the northwestern Atlantic. Growth rates are similar after the fourth year. Length at maturity (about 3.10 m for males and 3.15 m to 3.20 m for females) is about the same in both areas, but since the Gulf individuals grow faster, they mature at an earlier age. Gulf males are mature at 7.8 to 8.0 years, females at 8.8 years, whereas northwestern Atlantic males mature at more than 10.1 years and females at more than 9.0 years of age.

In one study, the ages of tiger sharks were determined by the vertebral ring (annulus) method. The largest male and female were each 3.81 m long, with ages of 15.1 and 16.1 years respectively. By extrapolating age based on lengths of specimens whose ages had been determined, larger specimens up to 4.5 m long could well be as old as 25 years. Theoretically, the largest tiger sharks may attain ages of up to 50 years.

One of the largest sharks, the tiger reputedly but doubtfully grows to the gigantic size of 9.1 m (30 ft); however, a photograph of a supposedly 7.4 m female from Indochina (Viet Nam?) is possibly the longest documented individual (also the heaviest at a reported 3,110 kg). Individuals over 5.5 m (18 ft) are rare; most adults captured are under 4.3 m. Tiger sharks weighing over 375 kg are commonly reported, but difficult to verify. Certainly the species attains weights in excess of 500 kg.

Although this species is large and dangerous, it affords great sport for the angler and is one of the seven types of sharks designated as a game fish by the International Game Fish Association. It is an economically important fish, especially in the West Indies, where it is eaten fresh, salted, or smoked, and where its skin is processed for leather and other products. Its teeth are commonly mounted as pendants on necklaces.

Further readings
Branstetter, S., et al., 1987; Casey, J. G., et al., (1) 1980, (2) 1981, (3) 1982, (4) 1983, (5) 1984, (6) 1985, (8) 1987; Clark, E., and K. v. Schmitt, 1965; Gudger, E. W., 1949; Kauffman, D. E., 1950; Springer, S., 1963.

Reef Sharks—Blacktip, Gray, and Whitetip

Why is a relatively small shark, the blacktip reef (adults measure less than 1.6 m) included among the supersharks? Because size is no measure of this shark's ferocity. The blacktip reef, *Carcharhinus melanopterus* (plate 9d) may be the only really dangerous species of small, shallow-water shark, although some scientists argue that the gray reef shark is even more belligerent. The blacktip reef occurs singly or in small groups, is abundant, and is the species most likely to be encountered by divers and waders in the shallow waters of the tropical Indo-Pacific. It is easily recognized by its bright blacktipped fins, the black first dorsal fin being most noticeable because of a contrasting white band beneath it. Although divers often regard the blacktip as timid, spearfishermen and waders can testify to its aggressiveness if bait or other food is available. It does not hesitate to approach and assault the legs of splashing waders, possibly mistaking them for struggling prey. It even enters water only a few inches deep with its back and tail fin exposed as it wriggles over the sand flats. Where fishermen are cleaning their catch or where there are lines of speared fish, it will attack. To avoid the blacktip, Marshall Islanders preferentially swim rather than wade when crossing atoll passes.

Despite its abundance and wide distribution throughout the tropical-subtropical western and central Pacific Ocean, Indian Ocean, Red Sea, and eastern Mediterranean (probably reached after the opening of the Suez Canal), relatively little is known about its biology. The species is believed to enter the brackish lakes and estuaries of Madagascar and freshwater in Malaysia, but probably does not remain in such environments very long. The blacktip reef shark eats small fishes, squid, octopus, and shrimp, but scant information is available on relative amounts of food or how long it may survive without eating. In the past it was not considered to be a good aquarium species because of difficulty in transporting it from the wild and because it would not eat in captivity. Three small blacktips survived for 2 months in captivity without eating, and a fourth individual lasted for 3 months, possibly feeding on its aquarium mates. One shark lived in an aquarium for 2 years, fed on pieces of fish. During this time, it grew from 48 to 94 cm (19 to 37 in) in length. More recently, however, new techniques in transporting and feeding blacktips have made them a desirable aquarium species.

In the wild, the hunter may become the hunted. Small groupers are part of the blacktip's diet, but the stomach contents of a large 36 kg (80 lb) grouper contained a 46 cm (18 in) blacktip.

Blacktip reef sharks are viviparous. It is not known exactly when they mate, although one report from the Red Sea cites early summer. Males reach maturity at 91 cm to 100 cm, females from 96 cm to 112 cm. Some males have been observed to follow females as if attracted by a female scent or pheromone, but this is another area where there are few facts and much conjecture. Females bear two to five young, about 33 cm to 52 cm long, reportedly from January to June, after a gestation period of perhaps 16 months.

Elevating the blacktip reef shark to the status of supershark invites comparison with two other sharks common throughout the coral atolls of the tropical Pacific: the gray reef shark, *Carcharhinus amblyrhynchos* (plate 4b) and the whitetip reef, *Triaenodon obesus* (plate 5a).

Some argue that the gray reef shark is more aggressive and equally if not more dangerous than the blacktip reef. The gray reef shark has attacked divers and small manned submersibles in situations when bait was offered and, strangely enough, even more readily without bait. When approached too closely or startled, these sharks will perform a series of threat postures, a distinct type of swimming display consisting of exaggerated and jerky movements (figure 52). These movements become more intense as the diver or submersible approaches rapidly or corners the shark. The shark arches its back, raises its snout (sometimes with the mouth open), lowers its pectoral fins, and in close encounters, it will swim in figure-8

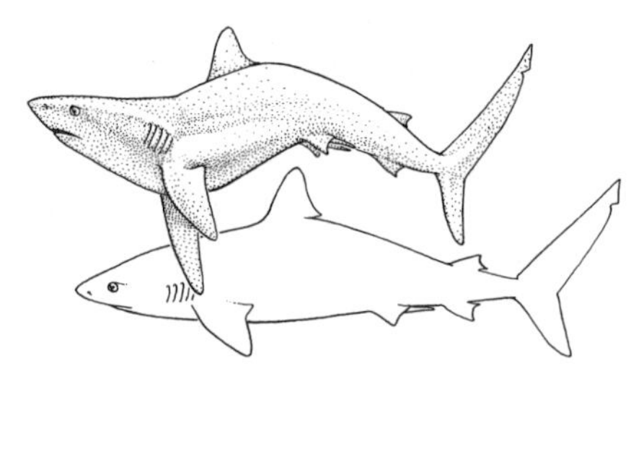

Figure 52. Gray reef shark, Carcharhinus amblyrhynchos, *threat postures superimposed on non-threat postures. (After Johnson and Nelson, 1973)*

loops. If these aggressive postures are not heeded, the shark will relentlessly pursue and attack the intruder. This behavior has been documented in experiments at Eniwetok in 1977 and 1978, using a vehicle called the Shark Observation Submersible. Why gray reef sharks behave this way is not certain. Results from the Eniwetok studies suggest the sharks are reacting defensively, as if man and submarine were predators rather than invaders of their territory or potential food items. What enemies might the gray reef shark normally encounter at Eniwetok? Both tiger sharks and Galapagos sharks, which may be larger than the gray reef, are known to occur in the area and are also known to eat other sharks. Furthermore, a young gray reef has been found in the stomach of a silvertip shark (not to be confused with the whitetip reef shark) from Eniwetok; however, it is not known how much predation occurs.

At Eniwetok, the blacktip, gray, and whitetip reef sharks inhabit different niches and different depths: the blacktip is found in the turbid sand flats from 0 m to 15 m, the gray in clear but deeper waters from the surface to 150 m, and sometimes as much as 1,000 m, and the whitetip hides in the crevices and caverns of the coral heads to about 30 m. In a recent study covering a 4-year period, gray reef sharks fitted with ultrasonic transmitters were tracked and found to occur in three daytime situations: near ocean-reef dropoffs in loose aggregations of as many as 20 individuals; just above level bottom areas, swimming in polarized schools of as many as 33 individuals; and over shallow reefs and lagoon pinnacles as lone individuals. Sharks tagged near the ocean reefs tended to range widely along the reef, while those tagged in the lagoon tended to return day after day to the tagging site. The reasons for these differences in behavior are unknown.

A loose aggregation of at least 12 gray reef sharks and two silvertips was also observed among schools of rainbow runners (a kind of jack fish). Surprisingly, the sharks made no attempt to eat the rainbow runners, which oddly rubbed their bodies against the sharks in what might have been a way to dislodge ectoparasites. When presented with newly speared fish, however, the sharks became excited and ate the bait.

Mature gray reef sharks are about the same size as adult blacktips, ranging from 1.4 m to 1.7 m and not exceeding 1.8 m. Reports of larger sizes, 2.3 m to 2.6 m, may be erroneous. The gray reef matures at about 7 years of age, males at 1.3 m to 1.4 m, females at 1.2 m to 1.4 m. They are viviparous, bearing one to six young, 45 cm to 60 cm long, after a gestation period of about 12 months. They, like the whitetip, are thought to live as long as 25 years. No longevity information is available for blacktip reef sharks.

Unlike the other two reef sharks, the whitetip reef shark is an unaggressive, slender shark that often rests in crevices and coral caves during the day. It usually stays close to the bottom to feed on reef fishes and extract octopuses from the holes in the reef. Even though found in company with other whitetips while resting in caves, these sharks do not swim together at night, when they are most active. They are inquisitive and often approach divers. With repeated experience, they learn that the activities of divers hold the promise of providing a meal of speared fish, and will leave their abodes at the drop of an anchor in anticipation.

Whitetip reef sharks are relatively small, normally not exceeding 1.8 m. In the only concentrated study of their biology, the largest of 231 specimens was a male 1.68 m in total length that weighed 23.6 kg (a slightly shorter specimen weighed 27.7 kg). This species may take 5 years to reach maturity, males maturing at 1.05 m and females at a somewhat larger size. One to five young, 52 cm to 60 cm long, are born, but the gestation period has not been determined. Whitetip sharks tagged, measured, and recaptured at Johnston Island after being at large for 6 months to a year were estimated to grow at a rate of 2.1 cm to 4.2 cm per year with a decreasing rate as maximum size was reached. Whitetips are thought to live as long as 25 years.

One of the mysteries surrounding the whitetip reef shark is how it attained its present distribution from the western Indian Ocean eastward all the way to the eastern Pacific coasts of Central America. This shark is always seen or caught on or near the bottom, never far from relatively shallow rocks and reefs, although there are unusual reports of captures at depths of about 110 m off Madagascar and 330 m off Okinawa, all from fishing lines set on the bottom. There is no reason to believe that the species swims at the surface over deep water, and it has never been reported to accompany flotsam as many fishes do. Furthermore, the species is relatively sluggish, moving no more than a few kilometers from tagging sites in periods of as much as a year. To have reached the eastern Pacific, the whitetip reef shark would have had to traverse landless expanses of ocean several thousand kilometers wide and over 4,000 m deep. Could the present-day distribution be a relic of the far distant past when the Pacific was much smaller than today and chains of islands extended from one side to the other?

Further readings

McKibben, J. N., and D. R. Nelson, 1986; Nelson, D. R., 1977, 1981; Nelson, D. R., et al., 1986; Randall, J. E., 1977; Randall, J. E., and G. S. Helfman, 1973; Russo, R., 1984.

PART THREE

Shark Attack

HOW SERIOUS IS THE THREAT?

Sharks do attack humans and are a potential threat to the life of anyone who spends time in the water. In recent years, the number of shark attacks may have increased, as some believe. If so, perhaps the increase is due to the increase in number of people engaging in water sports. As a result of the Marine Mammal Protection Act of 1973, there may also have been an increase in the populations of some species of sharks in response to increased populations of the marine mammals on which the sharks feed. On the other hand, the number of attacks may only be a manifestation of our interest. That is, the number of attacks may not have increased, but the proportion of attacks being reported and officially documented has increased. Whatever the circumstances, however, the threat of attack is not great. The dread of being eaten alive by a beast of the sea is far out of proportion to the number of attacks that occur. Less than 100 attacks are reported worldwide each year, and no more than 30 are fatal. To put this in perspective, over 50 fatal snakebites occur annually in Brazil alone. Many divers rarely catch a glimpse of a shark, and photographers may spend months or even years trying to attract one of the dangerous species within camera range. The sharks that do attack are mostly large individuals, between 2 m and 8 m in length, and more than half of these are members of the family Carcharhinidae, the requiem sharks.

WHICH SHARKS ARE DANGEROUS?

The four most dangerous sharks, which are those that always pose a danger to humans and have a record of numerous attacks with several fatalities, are the white shark, tiger shark, bull shark, and oceanic whitetip shark (not to be confused with the reef whitetip). These sharks are large, eat prey comparable in size to humans, and with the exception of the oceanic whitetip, visit shallow coastal areas where bathers are present. The oceanic whitetip is most often found well offshore in deep water, but occasionally ventures close to shore where the shelf is narrow. It is not a species to be taken lightly even though it has been implicated definitely in only one attack on man (nonfatal), almost 30 years ago. The oceanic whitetip is highly suspect and may well be responsible for many undocumented attacks on survivors of sea disasters.

Seventeen other species also considered dangerous, but with fewer numbers of recorded attacks and far fewer fatalities, are: the spotted and tasseled wobbegongs, nurse shark, spitting shark, sand tiger, shortfin mako, gray reef shark, narrowtooth shark, spinner shark, blacktip shark, blacktip reef shark, dusky shark, reef shark, sicklefin lemon shark, lemon shark, and possibly the Galapagos shark and blue shark. Several of these are unaggressive in situations where they meet divers, but will retaliate vigorously if they are disturbed, speared, or provoked. Blue sharks are rarely found in shallow areas, but because they travel in groups, unlike the other species, there is the potential for feeding frenzies to develop.

Four species, the Pacific angel shark, Atlantic angel shark, leopard shark, and whitetip reef shark, have bitten people, but ordinarily pose little danger because they are small or generally timid and have small teeth. Of greater danger than being bitten by the whitetip reef shark is the possibility of ciguatera poisoning from eating the liver of this shark, a favorite food item of the Gilbert Islanders.

Because of their massive size, the basking shark and whale shark should instill caution. Both these species are rather sluggish and lack sharp teeth, and normally pose no threat, even to divers intent on hitching a ride. When harpooned, however, the basking shark may ram boats.

Perhaps 40 other species are suspected of attacking people or are potentially harmful. These added to those known to be dangerous add up to less than 20 percent of all the known species of sharks.

WHERE DO ATTACKS OCCUR?

It has been reported that most shark attacks occur on sunny days in nearshore waters 20 degrees Celsius or warmer. This is probably true because the greatest numbers of swimmers prefer such conditions. But shark attacks also occur in deep waters and those at least as cool as 10 degrees Celsius. The white shark, in particular, normally patrols cold waters such as those off central and northern

California and Oregon. The "red triangle," from San Francisco's Golden Gate, about 40 km (25 miles) out to the Farallon Islands and north to Tomales Point, is the site of large sea lion and seal rookeries and is one of the most dangerous areas in which to surf or spearfish. The greatest number of white shark attacks off the North American coast occur near the Farallon Islands, but in general, for all species off North America, the greatest number of recorded shark attacks occur in Florida waters.

Several species are commonly encountered by divers on tropical reefs: nurse sharks, mackerel sharks of the family Lamnidae, hammerheads, and, most often, the requiem sharks of the family Carcharhinidae. More shark species may be found on tropical western Pacific reefs than on reefs elsewhere. The gray reef shark, blacktip reef shark, and whitetip reef shark are among the most common. Normally, there is little risk to divers unless the sharks are poked, pulled, harassed, or speared, or the warning signals or threat behavior exhibited by some species is ignored or the escape route of the sharks is blocked.

Prior to 1952, the beautiful bathing beaches of southeastern Australia and the Natal coast near Durban, South Africa, were the sites of frequent shark attacks. Since that time, offshore meshing, using a series of weighted gill nets set at overlapping intervals parallel to the beach and resting on the sea bottom, have practically eliminated shark attacks in those areas. The sharks seek the small fishes that are abundant in the rough, highly oxygenated, and turbid waters but frequently are caught in the nets and drown. The nets are regularly emptied of all sharks and other fish that have become entangled in them. One problem with the nets is that other animals, particularly marine mammals, occasionally are trapped in them and die.

WHY DO SHARKS ATTACK HUMANS?

Biologists are not sure. It is known that certain distress stimuli, such as the irregular, low-frequency sounds and odors emitted by speared fish or the sounds of splashing about in water, will attract sharks. But very little is known about when various species of sharks become hungry and how often they feed. A popular misconception is that all shark attacks are feeding forays. On the contrary, in his analysis of over 1,100 cases in the International Shark Attack File, David Baldridge estimated that from half to three-quarters may have been aggressive attacks that had nothing to do with feeding. The motivation for most unprovoked attacks in which the victim was not emitting distress signals may be hunger, but the examination of wounds on attack victims suggests other reasons as well.

Sharks are not infallible. They may mistake splashing waders for fish in trouble. They may mistake a surfer on his board for a seal. Some sharks attack humans once, then spit out their flesh and abandon them. The normal body secretions of a human even when not bleeding may be an attractant. Some sharks will test potential prey by "bumping" or ramming them before either attacking or departing.

In such encounters, victims often suffer painful abrasions and lacerations from the fins and rough denticles of the skin. Because the bites on some victims are similar to the courtship bites of males on females, even sex has been suggested as a motivating force. However, there is no proof that only male sharks perpetrate these attacks.

Some sharks will attack or display aggressiveness even though not directly provoked or excited by the presence of food. The victim may inadvertently intrude in the space a shark regards as his own, although the existence of such territorial behavior in sharks is still only a theory. Rather, as discussed in the section on the gray reef shark, some scientists believe that the series of threat postures performed when the shark is approached too closely by man or submersible is a reaction by the shark to a perceived danger of attack. If the shark's aggressive postures are not heeded, it will pursue and attack the interloper. At least five species of carcharhinid and sphyrnid sharks are known to exhibit some aspects of threat behavior.

WHERE ARE SHARK ATTACKS RECORDED?

In 1958, the U.S. Navy's Office of Naval Research convened a group of experts, known as the Shark Research Panel, with Dr. Perry W. Gilbert, then director of the Mote Marine Laboratory in Sarasota, Florida, as its chairman. Its function was to oversee the development of a shark research program, formulate the questions that were to be the basis on which shark research would be funded, and disseminate the results of their findings. The impetus for the creation of the panel was the realization that first, there was scant scientific information on the biology and behavior of sharks; and second, a serious morale problem had occurred during World War II, and could occur again, when servicemen realized they might have to abandon ship in shark-infested waters. These problems could only be addressed if an intensive research program were initiated.

In 1959, the International Shark Attack File was established by the Office of Naval Research for the purpose of providing a reliable data base for evaluating shark attacks. The nucleus of this file was a small cache of data collected over the years by the late Dr. Leonard Schultz, curator-in-charge of the Division of Fishes at the Smithsonian Institution. Cornell University, the Smithsonian Institution, and the American Institute of Biological Sciences worked jointly to collect and supply data to the file, which grew to 1,652 cases. In 1968, the data collected were sent to the Mote Marine Laboratory to be incorporated with their research data. The data were analyzed and presented in a report entitled, *Shark Attack against Man: A Program of Data Reduction and Analysis,* by Dr. H. David Baldridge, Mote Marine Laboratory, October 31, 1973. Mote Marine Laboratory provided the material, administrative, and technical support for the data reduction phase of preparing the report. The report attempts to explain how, when, where, and possibly why sharks attack, based on the cases analyzed. As a response to the public interest and inquiries regarding shark attacks, a popular version of the report, *Shark Attack*, also by H. David Baldridge, was published in 1974 by Berkeley Medallion Books of New York, and it remains a good source of information for the layman.

For some years all shark attack records were kept at Mote Marine Laboratory, but currently the American Elasmobranch Society has undertaken the responsibility for maintaining, expanding, and analyzing shark attack data. The files are under the direction of George Burgess, American Elasmobranch Society, Florida Museum of Natural History, Gainesville, Florida 32611.

ARE THERE ANY EFFECTIVE SHARK REPELLENTS?

No, neither chemical nor physical. The first chemical shark repellent widely used was developed in 1943, during World War II. It represented an effort to protect military personnel in the sea after their planes were downed or vessels sunk. "Shark Chaser," as it was known, was a water-soluble cake consisting of a mixture of 80 percent nigrosene dye and 20 percent copper acetate. In preliminary tests it appeared that the dye acted as a visual screen, and the copper acetate inhibited feeding responses by captive smooth dogfish sharks. In actual practice, "Shark Chaser" proved to be more of a morale booster than an effective shark repellent. Additional tests of the compound in 1963 showed that any effectiveness of "Shark Chaser" was due to the visual effect of the opaque dye, which, unfortunately, dispersed all too quickly in the open sea, rather than the copper acetate which had no effect as a feeding inhibitor.

Various devices have been tested as attack deterrents. The use of gill nets off the beaches in Australia and South Africa has been reasonably successful in decreasing the number of attacks by decreasing the size of the local shark population. But the use of gill nets is time consuming and expensive. Electrified and electronic barriers are being tried, but these are also expensive and can be dangerous to divers and marine animals other than sharks. A large foldable, plastic bag called the "Johnson Shark Screen" is a promising solution for survival of victims of sea or air disasters. The bag can be carried in a pocket, and at the time the victim finds himself in the water, unfolded, filled with water, and entered. A flotation ring at the top of the bag suspends it near the surface and supports the victim. Odors, sounds, electric fields, visual shapes, and movement normally attracting sharks are masked by the bag.

Among the protective devices developed for divers are the shark billy (a short pole to ward off the shark), the "bangstick" (a pole with an explosive cartridge that detonates on contact with the shark), a gas-injection dart (that bloats the shark and causes it to float), an electric dart, and a toxin syringe. Most of these weapons are more dangerous than the shark to the untrained user. Furthermore, the diver must be in position to see the shark he plans to engage. With some sharks, the time it takes for a drug to be absorbed, transported through the circulatory system, and incapacitate a shark is too slow to be safe. But these weapons seem to give some protection in specific situations, even though they may not always be relied upon.

In searching for a chemical repellent, several factors have to be considered: the repellent must be specific to sharks and it must have essentially no effect on the human user. It should be stable enough to be stored for a long time, be inexpensive, and effective in small amounts. The most promising naturally occurring chemical repellent derives from the shark-repellent properties of a flatfish from the Red Sea, the Moses sole (*Pardachirus marmoratus*). Dr. Eugenie Clark of the University of Maryland discovered that the Moses sole exudes a milky white secretion from glands along its dorsal and anal fins that has the effect of repelling sharks. The prime component of the secretion is the active toxin pardaxin, an extremely complex protein that affects the gills of sharks and is similar chemically to surfactants, detergents capable of emulsifying oils. Because pardaxin is too unstable and expensive to collect or synthesize, and because sufficient concentrations required for use in open water are difficult to determine, current research has focused on the production of a repellent based on industrial surfactants. Although surfactants have been found that are more effective than the freeze-dried toxic secretion of the Moses sole, the problem of swift dispersal in the water has not been solved. To date there is still no effective chemical repellent available in a small, handy cake or packet size for swimmers and divers. Current research efforts have diverged from the developing of chemical repellents to addressing the question of how to develop materials and products to protect undersea structures and devices from sharks.

Further readings

Baldridge, H. D., 1973, 1974; Brownlee, S., 1985; Coppleson, V. M., 1962; Davies, D. H., 1964; Gilbert, P. W., ed., 1963, Gilbert, P. W., 1977, 1984a, 1984b; Gruber, S. H., 1981a; Hodgson, L., 1984; Johnson, R. H., and D. R. Nelson, 1973; Nelson, D. R., 1981; Nelson, D. R., and R. H. Johnson, 1980; Nelson, D. R., et al., 1986; Wilson, R., and J. Q. Wilson, 1985; Zahuranec, B. J., ed., 1983.

HOW CAN SHARK ATTACKS BE AVOIDED?

The following advice, based on Perry W. Gilbert's chapter in *Sharks and Survival*, "Advice to Those Who Frequent, or Find Themselves in Shark-Infested Waters," chapter 18 in H. David Baldridge's *Shark Attack*, and information from Leonard J.V. Compagno's *Sharks of the World*, is rephrased and condensed. These are commonsense rules about where and when to swim or dive and how to avoid attracting sharks. They include some new warnings based on what is known of shark behavior.

1. Do not enter waters known to be frequented by large sharks.
2. Do not enter waters where there has been a recent shark attack.
3. Do not swim or dive alone. Always have a companion or stay in a group.

Most attacks occur on lone individuals. A companion can render assistance if close by.

4. People with open wounds, even slight cuts, and divers with injuries should stay out of the water or leave quickly if injured. Blood and human wastes may attract and excite sharks.

5. Swimming or diving in murky water should be avoided. The swimmer should be watchful in this situation.

6. Avoid swimming far offshore, in channels, or over dropoffs to deeper water—these are all areas where large sharks are most often encountered.

7. If groups of fish start to behave erratically or congregate in large numbers, leave the water.

8. Avoid entering the water if you are unevenly suntanned or wearing a swimsuit that is bright or contrasts with your skin. Low contrast is better. Dark-skinned swimmers should wear dark-colored swimsuits, and light-skinned ones should wear light blue or green. Do not wear colors such as yellow, orange, and red. Reverse countershading should also be avoided, for instance, a two-color suit that is dark on the front side and light on the back, like a floating dead fish. Wetsuit color should be chosen carefully so it does not mimic natural prey. The wearing of a black and white striped wet suit that mimics the coloration of small, venomous sea snakes, was reported by one ichthyologist to discourage sharks, but others have not had the same good experience. Indeed, some tropical Pacific and Indian Ocean sharks actually eat banded sea snakes. Although the gray reef sharks off the Australian coast appeared to be repelled by a diver in a black and white striped suit, blue sharks off California and gray reef, blacktip reef, and whitetip reef sharks in the Pacific were not.

9. Do not swim with dogs or other pets, or surf near seal and sea lion colonies.

10. Scan the water for sharks before jumping or diving from a boat.

11. Be alert for an entrapped shark when swimming in tide pools, bays, or small lagoons that have been cut off from the sea at low tide.

12. Avoid swimming or diving at dawn, dusk, or at night when many species of sharks feed more actively.

13. Do not provoke or molest a shark, spearing or riding it, or pulling its tail. Even small sharks can inflict painful or serious wounds.

14. Surfers should be aware that the risk of shark attack appears to be greater with the use of the newer short bellyboards than with the older larger surfboards. The overall silhouette of a surfer's arms and legs hanging from a bellyboard might be mistaken for the shape of a seal by a dangerous shark such as the white shark.

15. Avoid spearfishing in the same waters for an extended period. Erratic swimming movements and the body fluids of wounded fish increase chances of attracting sharks. Do not carry speared fish on or near your body; remove them from the water immediately.

16. Divers should leave the water as soon as possible if a large shark is sighted. Do not panic. Avoid erratic movements; swim smoothly to reach the shore or boat. If wearing scuba, swim along the bottom, where rocks or corals might provide protection, to below the boat before surfacing. Submerged divers should watch

sharks carefully to determine the best time to leave the water, or counter an attack should one occur. When menaced, back off, keeping your movements calm. Any rigid or jerky movements by a shark may represent a threat display, and the diver should retreat at once.

17. Shark cages or small sharkproof submersibles may be necessary for divers who must work in a specific area where they are harassed by dangerous sharks. Custom-made stainless steel, chain-mail suits have been made that provide limited protection from the bites of small sharks, but these suits are inconvenient to wear, generally unavailable, and too expensive to be practical.

Even if you do not or cannot conform with any or all the above recommendations, it may be reassuring to remember that the likelihood of being attacked by a shark is slight. On the other hand, conforming with all the recommendations is no guarantee that a shark attack will be avoided. If one desires absolute safety from shark attacks in marine waters, we recommend bathing in the Black Sea (bordered by Turkey, Bulgaria, Romania, and the USSR), which is about half the saltiness of normal seawater, its appendix, the Sea of Azov, or the landlocked Caspian Sea (both USSR), which is even less salty. No shark attacks have been reported from these waters, and no sharks exist in the Caspian Sea.

What to Do if a Shark Attack Appears Imminent

1. Keep calm and use any weapons you may have to fend off the shark. Divers should carry a tool or object such as a shark billy (a long club to fend off or push off a shark), a speargun, abalone iron, or the like.

2. Avoid injuring the shark unless all else fails. An injury may make the shark more aggressive. Spearguns, powerheads (bang sticks), or gas guns should be used as a last resort, as they may only excite the shark or attract other sharks to the area.

3. Sometimes an attack may be aborted by shouting underwater, blowing bubbles, or poking the shark, even with your hands, in its eyes or gills. At other times these actions have no effect and even provoke more aggressive behavior.

4. If you are bitten, make immediate efforts to stop the bleeding, even before leaving the water. Most victims die from loss of blood rather than from loss of body parts.

PART FOUR

Sharks and Us

OF WHAT USE ARE SHARKS?

It is not true in the United States, but the main and most important use of sharks worldwide has always been for food. Sharks have been salted, dried, barbecued, marinated, minced, ground into meal, and converted into delicacies such as shark-fin soup, and hakall, an Icelandic specialty made from the Greenland shark, which is poisonous if not properly prepared. Although the United States has not wholeheartedly embraced foods like mako steak or shark teriyaki, recipes for such items are appearing in cookbooks published in East and West Coast beach communities. A recent edition of the popular *Joy of Cooking* suggests preparing shark as you would fish fillets. In some nations, particularly along the coasts of Asia and in the South Seas, shark meat has always formed a normal part of the dietary protein. Because of the increase in human populations throughout the world, shark meat will almost certainly become more common as a source of protein.

In the United States most people know very little about the edibility of sharks, how they taste, and how they should be prepared before cooking. Most people have an inbred aversion to or bias against sharks as food, and most fishermen discard the sharks they catch. Except for the Greenland shark, which occurs from Cape Cod northward, the flesh, excluding the viscera, of all sharks with five gill openings caught along the coasts of the United States is edible. The easily recognized sixgill and sharpnose sevengill sharks are reported to be mildly toxic, but the flesh, after it is rinsed several times with water, appears to be innocuous.

If shark flesh is cooked fresh or within a few hours of being caught, no special preparations, other than rinsing the flesh in fresh water, are necessary. If the meat is to be held for longer periods, the shark must be bled and the flesh rinsed and rapidly chilled so it will not spoil. Normally, shark flesh and blood contain relatively high concentrations of two nitrogen compounds, urea and trimethylamine oxide. These substances pose no problem in fresh shark meat, but on standing they produce ammonia and other noxious odors.

During the 1940s, shark-liver oil replaced cod-liver oil as a major source of vitamin A, but after 1950 this natural source became obsolete with the advent of less expensive synthetic products.

Currently, several ongoing medical research projects are using shark parts or extracts, but so far all are in the developmental stage. For example, recent studies of shark cartilage may provide hope for cancer victims. Extracts of a substance occurring in the cartilage have been found to inhibit the growth of tumors implanted in laboratory animals. Initially, the studies were conducted using calf cartilage; however, only a small amount of cartilage is present in mammals. So basking shark cartilage, which constitutes about 6 percent of the shark's weight, was substituted. Extracts from the cartilage significantly slowed the growth of tumors implanted into the eyes of rabbits by inhibiting the growth of blood vessels leading to the tumors. Although further studies are needed, the use of these cartilage extracts may provide a means of controlling certain cancers. The substance found in shark cartilage may also be responsible for the rarity of tumors in sharks. Chondroiten, also derived from shark cartilage, is being tested for making artificial skin for burn patients. Shark corneas have been used as replacements for human corneas, and heparin-like antibloodclotting agents have been extracted from certain sharks such as the spiny dogfish.

All other uses of sharks fill a minor place in the economy. Live animals are exhibited in public aquariums or used as test animals in research; whole dead specimens are used in training zoology students in anatomy; processed carcasses are used in making fertilizer or feed for livestock; skin is used for leather products, such as shoes, wallets, and belts; teeth are used for ornaments and, in primitive societies, as parts of weapons, and jaws are mounted as curios; shark liver oils (pristane, squalene) are used for cosmetics and special lubricants.

Sharks contribute importantly to our lives, but considering how much fear there is of shark attacks and interest in preventing them, one might think that getting rid of sharks would be a priority occupation of mankind. Many of those who earn their livelihood at beachside tourist resorts might agree. Nowadays, it is common to read reports in the newspapers of large sharks caught and killed by fisherman simply for the sport. What is unusual is that we are now hearing many people express their displeasure with such irresponsible behavior. It is possible that large marine predators, like terrestrial predators, can be eliminated by man's activities. But what would the world be like without sharks? Would something else evolve to replace them, or would their loss affect disastrously the ecological balance of the seas? We really cannot predict what would happen, but can we risk finding out? We think not.

There are definite indications that we are irrevocably destroying our natural heritage, even at the molecular level. We should be alert to pleas from biologists to conserve the DNA (raw material of genetic variability) of potentially endangered species because of its possible use in genetic engineering or for the clues it provides to the evolutionary history of all organisms. Furthermore, the elimination of each shark species equals the loss of a potentially important source of food. Finally, there is the most fundamental philosophical question: do humans have the right to extinguish any species?

Further readings
Gordievskaya, V. S., 1973; Lee, A., and R. Langer, 1983; Otwell, W. S., et al., 1985; Ronsivalli, L. J., 1978; Seymour, H. H., and K. S. Danberg, n.d.

HOW CAN I BECOME A SHARK SPECIALIST?

With many years of schooling, serious application, and determination.

The study of sharks is a branch of ichthyology, the study of fishes, which, in turn, is a branch of zoology or biology. Many scientists who study sharks do not consider themselves ichthyologists. For instance, one anatomist we know studies vertebrate brain structure. He has published many papers on the brains of fishes, including those of sharks, as well as all other vertebrate groups. He describes himself as a comparative vertebrate anatomist rather than an ichthyologist or elasmobranchologist. There are ichthyologists, including some of the most prominent shark experts, that study other fishes as much as they do sharks. They probably describe themselves primarily as ichthyologists, as do those ichthyologists who restrict their studies to sharks. Whatever -ologist title one assumes is pretty much a personal decision. Professionals in the field are interested mainly in how good you are at what you do, not what you call yourself.

If you are serious about studying sharks, especially if you expect to make a living doing it, you are going to have to earn a college degree, probably two or three degrees. No university, marine laboratory, government agency, or museum will employ a person without a college degree as a research scientist. Today, an M.S., and more commonly a Ph.D., is mandatory for gaining employment as a professional researcher in any branch of science.

Although many universities offer coursework in ichthyology, none offers a course in elasmobranchology, or even an undergraduate degree in ichthyology. The first step along the way to a career in ichthyology is to obtain an undergraduate degree with a major in the biological sciences. Most universities have their own requirements for undergraduate specialization in this subject area, but courses in general biology, zoology, physiology, comparative anatomy (almost essential, but rarely offered nowadays), embryology, genetics, and molecular biology are all appropriate. For any area of biology, a firm grounding in the basic sciences (chemistry, physics, and mathematics through calculus) are extremely important. Facility in statistics and with foreign languages (many shark studies are published in German, French, Russian, and Spanish, but not in Latin) and computers vary from helpful to mandatory. One of the most important tools one can acquire for any field of specialization is competency in English composition. Scientists advance their careers by the research they publish, and the ability to write clearly is a requirement for publication. Besides, clear writing and clear thinking go hand in hand.

There are so many courses in which one needs grounding, that the first year of graduate work is often spent taking courses one missed as an undergraduate.

Getting into graduate school is difficult. First of all, a B average or higher, especially in the sciences, is an almost universal requirement for gaining admission, as is a high score on the Graduate Record Examination. Particularly good students are sometimes offered graduate assistantships, teaching fellowships, and scholarships that help pay the bills. Finding the right graduate school will require some searching. If you have not wavered in your desire to study sharks by the beginning of your senior year, you should seek out a university that is associated with a marine laboratory (for example, University of Miami, University of Rhode Island, University of Washington, University of California at San Diego, Texas A & M University, and several others just as good). There will probably be an ichthyologist on the staff that you can write or visit to discuss your interests and plans. Every university professor is looking for good graduate students, and you will receive a cordial response to your inquiries.

Almost any special talent or knowledge you have can be an advantage to you as a scientist. A knowledge of photography, boat handling, motor repair, carpentry, scuba, electronics, drafting, illustration, fishing methods, data processing, and so forth, will be important at some or all stages of your career.

Some other things you can do may help. You can join the American Elasmobranch Society, an organization dedicated to the increase and diffusion of knowledge about sharks and other elasmobranch fishes. Anyone can become a member, even a high school student or a botanist. To do so is to support the aims of the society. The AES periodically distributes a newsletter to members, and, although it does not have its own journal, it anticipates that it will in the near future. For information about the AES, write Dr. Richard Lund, Biology Department, Adelphi University, Garden City, New York 11530. The AES holds an annual meeting (usually in the summer months when professors are free). Attend a meeting (even if you do not join), and listen to the papers that are being given; many are of broad general interest. Introduce yourself to the people giving the papers and indicate your interests.

If you can afford to volunteer your services during the summer, as a research assistant willing to do even the most menial tasks, you will gain far more than just the experience of working on a research project. You might become indispensable; *every* serious, hardworking volunteer in the field of science that we have known (and there have been many) has been offered a paying job, sooner or later, by the recipient organization. Jobs in science are difficult to obtain, but volunteers get to the head of the line quickest.

The study of sharks attracts a varied group of people, including a rather large proportion that seek the public limelight. Good show people are good for all science, as long as the information they present is accurate and the descriptive adjectives kept in bounds.

The American Society of Ichthyologists and Herpetologists, Inc., (c/o Dr. Carter R. Gilbert, Florida Museum of Natural History, Gainesville, Florida, 32611) will send a brochure, *Career Opportunities in Ichthyology,* on request, that also contains useful information.

Where Are Shark Specialists Employed?

At universities, museums, marine laboratories, and public aquariums; by private, federal, and state organizations. In the United States, most shark specialists are employed by universities and museums. Those at universities usually have both teaching and research as major responsibilities; whereas those at museums have research, care of collections, and exhibits as their major responsibilities (many museum curators also teach). University shark specialists and those at marine laboratories usually work on some aspect of the life history or biology of sharks. Museum workers are mostly concerned with studies on the evolution and classification of sharks. Some shark specialists are administrators in their respective organizations, always a disconcerting possibility the better known and more highly regarded one becomes. Salaries vary widely, from barely livable for the starting professional, to highly adequate for professors, senior government scientists, and aquarium directors. Few, if any, get rich, but many have interesting and exciting careers.

WHAT IS LEFT TO LEARN ABOUT SHARKS?

A lot! With 368 recognized species of sharks to study, an endless number of research problems are waiting to be solved. From the information provided in this book you can see that just the answer to the question, "how long do sharks live?" is still not possible to give with assurance. Even among the best-studied sharks, there are problems in age determination and disagreements among the specialists as to what their data really mean. Yet, age determination is one of the most important pieces of information needed to understand the biology of sharks. Even the exact number of shark species is known only approximately, with new species being described every year. Some of these new species are based on newly acquired specimens, but some changes in the number of species are the result of a reexamination and reinterpretation of previously available specimens. Our knowledge of shark behavior is still in its infancy; very few shark species have been the subject of controlled observation and experimentation.

You can be certain that many more books about sharks will be written as new information increases our understanding; quite possibly, the future authors are among our readers.

PART FIVE

Appendix 1

CLASSIFICATION OF SHARKS[1]

Class Chondrichthyes
Subclass Elasmobranchii
Superorder: Squalomorphii
Order: Hexanchiformes (cow sharks and frill sharks)
Family: Hexanchidae (cow sharks, sixgills, sevengills)
Genus: *Heptranchias*
H. perlo
Hexanchus
H. griseus
H. vitulus
Notorynchus
N. cepedianus
Family: Chlamydoselachidae (frill sharks)
Genus: *Chlamydoselachus*
C. anguineus
Order: Squaliformes (dogfish sharks)
Family: Echinorhinidae (bramble sharks)
Genus: *Echinorhinus*
E. brucus
E. cookei
Family: Squalidae (dogfish sharks)
Genus: *Aculeola*
A. nigra
Centrophorus
C. acus
C. granulosus
C. harrissoni
C. lusitanicus
C. moluccensis
C. niaukang
C. squamosus
C. tesselatus
C. uyato
Centroscyllium
C. fabricii
C. granulatum
C. kamoharai
C. nigrum
C. ornatum
C. ritteri
C. sheikoi[17]
Centroscymnus
C. coelolepis
C. crepidater
C. cryptacanthus
C. owstoni

Cirrhigaleus
C. barbifer
Dalatias
D. licha
Deania
D. calcea
D. histricosa
D. profundorum
D. quadrispinosum
Etmopterus
E. baxteri
E. brachyurus
E. bullisi
E. carteri[2]
E. decacuspidatus
E. frontimaculatus[17]
E. gracilispinis
E. granulosus
E. hillianus
E. lucifer
E. molleri[3]
E. perryi[2]
E. polli
E. princeps
E. pusillus
E. schmidti[17]
E. schultzi
E. sentosus
E. spinax
E. unicolor
E. villosus
E. virens
Euprotomicroides
E. zantedeschia
Euprotomicrus
E. bispinatus
Heteroscymnoides
H. marleyi
Isistius
I. brasiliensis
I. plutodus
Mollisquama[4]
M. parini
Scymnodalatias
S. albicauda[4]
S. sherwoodi
Scymnodon[4]
S. ichiharai
S. macracanthus
S. obscurus
S. plunketi
S. ringens
Somniosus
S. microcephalus
S. pacificus
S. rostratus
Squaliolus
S. aliae[18]
S. laticaudus
Squalus[5]
S. acanthias
S. asper
S. blainvillei
S. cubensis
S. japonicus
S. megalops
S. melanurus
S. mitsukurii
S. rancureli
Zameus[4]
Z. squamulosis
Family: Oxynotidae (rough sharks)
Genus: *Oxynotus*
O. bruniensis
O. caribbaeus
O. centrina
O. japonicus[6]
O. paradoxus
Order: Pristiophoriformes (sawsharks)
Family: Pristiophoridae (sawsharks)
Genus: *Pliotrema*
P. warreni
Pristiophorus
P. cirratus
P. japonicus
P. nudipinnis
P. schroederi
Superorder: Squatinomorphii
Order: Squatiniformes (angel sharks)
Family: Squatinidae (angel sharks, sand devils)
Genus: *Squatina*
S. aculeata

S. africana
S. argentina
S. australis
S. californica
S. dumeril
S. formosa
S. japonica
S. nebulosa
S. oculata
S. squatina
S. tergocellata
S. tergocellatoides

Superorder: Galeomorphii
Order: Heterodontiformes (bullhead)
Family: Heterodontidae (bullhead sharks, horn sharks)
Genus: *Heterodontus*
H. francisci
H. galeatus
H. japonicus
H. mexicanus
H. portusjacksoni
H. quoyi
H. ramalheira
H. zebra

Order: Orectolobiformes (carpet sharks)
Family: Parascyllidae (collared carpet sharks)
Genus: *Cirroscyllium*
C. expolitum
C. formosanum
C. japonicum
Parascyllium
P. collare
P. ferrugineum
P. multimaculatum
P. variolatum

Family: Brachaeluridae (blind sharks)
Genus: *Brachaelurus*[7]
B. waddi

Family: Orectolobidae (wobbegongs)
Genus: *Eucrossorhinus*
E. dasypogon
Orectolobus
O. japonicus
O. maculatus
O. ornatus
O. wardi
Sutorectus
S. tentaculatus

Family: Hemiscyllidae (bamboo sharks)
Genus: *Chiloscyllium*[8]
C. arabicum[9]
C. burmensis
C. griseum
C. hasselti
C. indicum
C. plagiosum
C. punctatum
Hemiscyllium
H. freycineti
H. hallstromi
H. ocellatum
H. strahani
H. trispeculare

Family: Rhincodontidae (whale sharks, zebra sharks, nurse sharks)[10]
Genus: *Ginglymostoma*
G. cirratum
Pseudoginglymostoma
P. brevicaudatum
Nebrius
N. ferrugineus
Stegostoma
S. fasciatum
Rhincodon
R. typus

Order: Lamniformes (mackerel sharks)
Family: Odontaspididae (sand tiger sharks)
Genus: *Eugomphodus*
E. taurus
E. tricuspidatus
Odontaspis
O. ferox
O. noronhai

Family: Mitsukurinidae (goblin sharks)
Genus: *Mitsukurina*
M. owstoni

Family: Pseudocarchariidae (crocodile sharks)
Genus: *Pseudocarcharias*
P. kamoharai
Family: Megachasmidae (megamouth sharks)
Genus: *Megachasma*
M. pelagios
Family: Alopiidae (thresher sharks)
Genus: *Alopias*
A. pelagicus
A. superciliosus
A. vulpinus
Family: Cetorhinidae (basking sharks)
Genus: *Cetorhinus*
C. maximus
Family: Lamnidae (mackerel sharks, porbeagles, white sharks)
Genus: *Carcharodon*
C. carcharias
Isurus
I. oxyrinchus
I. paucus
Lamna
L. ditropis
L. nasus
Order: Carcharhiniformes (ground sharks)
Family: Scyliorhinidae (cat sharks)
Genus: *Apristurus*[11]
A. abbreviatus
A. acanutus
A. atlanticus
A. brevicaudatus
A. brunneus
A. canutus
A. fedorovi
A. gibbosus
A. herklotsi
A. indicus
A. investigatoris
A. japonicus
A. kampae
A. laurussoni
A. longianalis
A. longicaudatus
A. longicephalus
A. macrorhynchus
A. macrostomus
A. maderensis
A. manis
A. microps
A. micropterygeus
A. nasutus
A. parvipinnis
A. pinguis
A. platyrhynchus
A. profundorum
A. riveri
A. saldanha
A. sibogae
A. sinensis
A. spongiceps
A. stenseni
A. verweyi
A. xenolepis
Asymbolus
A. analis
A. vincenti
Atelomycterus
A. macleayi
A. marmoratus
Aulohalaelurus
A. labiosus
Cephaloscyllium
C. fasciatum
C. isabellum
C. laticeps
C. nascione
C. silasi
C. sufflans
C. ventriosum
Cephalurus
C. cephalus
Galeus
G. arae
G. boardmani
G. eastmani
G. longirostris[12]
G. melastomus
G. murinus
G. nipponensis
G. piperatus
G. polli

G. sauteri
G. schultzi
Halaelurus
H. alcocki
H. boesemani
H. buergeri
H. canescens
H. clevai[19]
H. dawsoni
H. hispidus
H. immaculatus
H. lineatus
H. lutarius
H. natalensis
H. quagga
Haploblepharus
H. edwardsii
H. fuscus
H. pictus
Holohalaelurus
H. punctatus
H. regani
Parmaturus
P. campechiensis
P. macmillani[20]
P. melanobranchius
P. pilosus
P. xaniurus
Pentanchus[11]
P. profundicolus
Poroderma
P. africanum
P. marleyi
P. pantherinum
Schroederichthys
S. bivius
S. chilensis
S. maculatus
S. tenuis
Scyliorhinus
S. besnardi
S. boa
S. canicula
S. capensis
S. cervigoni
S. garmani
S. haeckelii
S. hesperius
S. meadi
S. retifer
S. stellaris
S. torazame
S. torrei

Family: Proscylliidae (finback cat sharks)
Genus: *Ctenacis*
C. fehlmanni
Eridacnis
E. barbouri
E. radcliffei
E. sinuans
Gollum
G. attenuatus
Proscyllium
P. habereri

Family: Pseudotriakidae (false cat sharks)
Genus: *Pseudotriakis*
P. microdon

Family: Leptochariidae (barbeled houndsharks)
Genus: *Leptocharias*
L. smithii

Family: Triakidae (houndsharks)
Genus: *Furgaleus*
F. macki
Galeorhinus
G. galeus
Gogolia
G. filewoodi
Hemitriakis
H. japanica
Hypogaleus
H. hyugaensis
Iago
I. garricki
I. omanensis
Mustelus
M. antarcticus
M. asterias
M. californicus
M. canis
M. dorsalis
M. fasciatus
M. griseus
M. henlei
M. higmani
M. lenticulatus

M. lunulatus
M. manazo
M. mento
M. mosis
M. mustelus
M. norrisi
M. palumbes
M. punctulatus
M. schmitti
M. whitneyi
Scylliogaleus
S. quecketti
Triakis
T. acutipinna
T. maculata
T. megalopterus
T. scyllium
T. semifasciata

Family: Hemigaleidae (weasel sharks)
Genus: *Chaenogaleus*
C. macrostoma
Hemigaleus
H. microstoma
Hemipristis
H. elongata
Paragaleus
P. leucolomatus[13]
P. pectoralis
P. tengi

Family: Carcharhinidae (requiem sharks)
Genus: *Carcharhinus*[14]
C. acronotus
C. albimarginatus
C. altimus
C. amblyrhynchos
C. amblyrhynchoides
C. amboinensis
C. borneensis
C. brachyurus
C. brevipinna
C. cautus
C. dussumieri
C. falciformis
C. fitzroyensis
C. galapagensis
C. hemiodon
C. isodon
C. leucas
C. limbatus
C. longimanus[15]
C. macloti
C. melanopterus
C. obscurus
C. perezi
C. plumbeus
C. porosus
C. sealei
C. signatus
C. sorrah
C. wheeleri
Galeocerdo
G. cuvier
Glyphis
G. gangeticus
G. glyphis
Isogomphodon
I. oxyrhynchus
Lamiopsis
L. temmincki
Loxodon
L. macrorhinus
Nasolamia
N. velox
Negaprion
N. acutidens
N. brevirostris
Prionace
P. glauca
Rhizoprionodon
R. acutus
R. lalandii
R. longurio
R. oligolinx
R. porosus
R. taylori
R. terraenovae
Scoliodon
S. laticaudus
Triaenodon
T. obesus

Family: Sphyrnidae (bonnethead sharks, hammerhead sharks, scoophead sharks)

Genus: *Eusphyra*
E. blochii
Sphyrna[16]
S. corona
S. lewini
S. media
S. mokarran
S. tiburo
S. tudes
S. zygaena

1. Unless otherwise indicated in these notes, classification is based on Leonard J. V. Compagno, 1977, Phyletic Relationships of Living Sharks and Rays, *American Zoologist,* 17(2): 303–322, and Sharks of the World, 1984, *FAO Species Catalog,* vol. 4, Part 1 and 2, Rome, Food and Agriculture Organization of the United Nations.

2. Stewart Springer and George H. Burgess, 1985, Two New Dwarf Dogsharks (*Etmopterus,* Squalidae), Found off the Caribbean Coast of Colombia. *Copeia,* no. 3, pp. 584–591.

3. Takashi Yamakawa et al., 1986, Review of the *Etmopterus lucifer* Group (Squalidae) in Japan. *In* T. Uyeno et al., eds., *Indo-Pacific Fish Biology: Proceedings of the Second International Conference on Indo-Pacific Fishes.* Tokyo: Ichthyological Society of Japan, pp. 197–207.

4. Toru Taniuchi and J. A. F. Garrick, 1986, A New Species of *Scymnodalatias* from the Southern Oceans, and Comments on Other Squaliform Sharks. *Japanese Journal of Ichthyology,* 33(2): 119–134.

5. N. A. Myagkov and V. V. Kondyurin, 1986, Dogfishes of the Genus *Squalus* from the Atlantic Ocean with Comparative Remarks on Members of the Genus from Other Regions. *Journals of Ichthyology, USSR,* 26(4): 560–575 [translated from the Russian]. The authors did not recognize *S. melanurus, S. mitsukurii,* and *S. rancureli,* and added *S. acutipinnus, S. fernandinus, S. lebruni,* and *S. probatovi.* N. V. Parin, 1987, Species of Spiny Dogfish of Genus *Squalus,* Living on Southeastern Pacific Ocean Seamounts, *Journal of Ichthyology, USSR,* 27(5): 43–50 [translated from the Russian] seriously questioned the validity of Myagkov and Kondyurin's actions.

6. K. Yano and M. Murofushi, 1985, A New Prickly Dogfish, *Oxynotus japonicus,* from Japan. *Japanese Journal of Ichthyology,* 32(2): 129–136.

7. Guido Dingerkus, 1986, Interrelationships of Orectolobiform Sharks (Chondrichthyes: Selachii). *In* T. Uyeno et al., eds., *Indo-Pacific Fish Biology: Proceedings of the Second International Conference on Indo-Pacific Fishes.* Tokyo: Ichthyological Society of Japan, pp. 227–245. Dingerkus places *Heteroscyllium* in the synonomy of *Brachaelurus* and *H. colcloughi* in the synonymy of *B. waddi.*

8. Guido Dingerkus and Terry C. DeFino, 1983, A Revision of the Orectolobiform Shark Family Hemiscyllidae (Chondrichthyes, Selachii). *Bulletin of the American Museum of Natural History,* vol. 176, article 1, pp. 1–94.

9. For discussion of the name *arabicum,* see John E. Randall, 1986, *Sharks of Arabia.* London: Immel Publishing.

10. Guido Dingerkus, 1986, Interrelationships of Orectolobiform Sharks (Chondrichthyes: Selachii). *In* T. Uyeno et al., eds., *Indo-Pacific Fish Biology: Proceedings of the Second International Conference on Indo-Pacific Fishes.* Tokyo: Ichthyological Society of Japan. Dingerkus (6) synonymized Ginglymostomatidae and Stegostomatidae with the Rhincodontidae, in which he included *Ginglymostoma, Stegostoma, Nebrius, Rhincodon,* and *Pseudoginglymostoma* (a new genus assigned to a species Compagno included in Ginglymostoma).

11. According to L. J. V. Compagno, it is probable that *Apristurus* is a junior synonym of *Pentanchus,* and that *A. herklotsi* is a junior synonym of *Pentanchus profundicolus.* Studies of these sharks are in progress by workers in New Zealand and Japan. The species list, which includes several recent studies in Russian and Chinese that were unavailable to us, was provided by Dr. J. A. F. Garrick.

12. H. Tachikawa and T. Taniuchi, 1987, *Galeus longirostris,* a New Species of the Sawtail Catshark from Japan. *Japanese Journal of Ichthyology,* 33(4): 352–359.

13. L. J. V. Compagno and Malcolm J. Smale, 1985, *Paragaleus leucolomatus,* a New Shark from South Africa, with Notes on the Systematics of Hemigaleid Sharks (Carcharhiniformes: Hemigaleidae). *Special Publication No.* 37, Grahamstown, South Africa: J.L.B. Smith Institute of Ichthyology, pp. 1–21.

14. J. A. F. Garrick, 1985, Additions to a Revision of the Shark Genus Carcharhinus: Synonomy of *Aprionodon* and *Hypoprion,* and Description of a New Species of *Carcharhinus. U.S. Dept. of Commerce, NOAA Technical Report* NMFS 34, 26 pp.

15. The oldest scientific name for this species is *Carcharhinus maou.* Although correct, this name has rarely been used. We hope the name will be suppressed to avoid confusion.

16. J. D. McEachran and B. Seret, 1987, Allocation of the Name *Sphyrna tudes* (Valenciennes, 1822) and Status of the Nominal Species *Sphyrna couardi* Cadenat, 1951 (Chondrichthyes, Sphyrnidae). *Cybium,* 11(1): 39–46. *S. couardi* is synonymized with *S. lewini.*

17. V. N. Dolganov, 1986, Description of New Species of Sharks of the Family Squalidae (Squaliformes) from the North-western Part of the Pacific Ocean with Remarks of Validity of *Etmopterus frontimaculatus. Zoologicheskii Zhurnal,* 65(1): 149–153. (In Russian with English summary.)

18. K. Sasaki and T. Uyeno, 1987, *Squaliolus aliae,* a Dalatiid Shark Distinct from *S. laticaudus. Japanese Journal of Ichthyology,* 34(3): 373–376.

19. B. Seret, 1987, *Halaelurus clevai,* sp. n., a New Species of Catshark (Scyliorhinidae) from off Madagascar, with Remarks on the Taxonomic Status of the Genera *Halaelurus* Gill and *Galeus* Rafinesque. *Special Publication No. 44,* Grahamstown, South Africa: J.L.B. Smith Institute of Ichthyology, pp. 1–28.

20. G. S. Hardy, 1985, A New Species of Catshark in the Genus *Parmaturus* Garman (*Scyliorhinidae*), from New Zealand. *New Zealand Journal of Zoology,* 12: 119–124.

Appendix 2

SELECTED LIST OF COMMON AND SCIENTIFIC NAMES OF SHARKS[1]

Atlantic angel shark—*Squatina dumeril*
Atlantic sharpnose shark—*Rhizoprionodon terraenovae*
Bahamas sawshark[2]—*Pristiophorus schroederi*
Basking shark—*Cetorhinus maximus*
Bigeye thresher—*Alopias superciliosus*
Bigeyed sixgill shark[2]—*Hexanchus vitulus*
Bignose shark—*Carcharhinus altimus*
Black dogfish—*Centroscyllium fabricii*
Blacknose shark—*Carcharhinus acronotus*
Blacktip reef shark[2]—*Carcharhinus melanopterus*
Blacktip shark—*Carcharhinus limbatus*
Blue shark—*Prionace glauca*
Bonnethead—*Sphyrna tiburo*
Bramble shark—*Echinorhinus brucus*
Broadband dogfish—*Etmopterus gracilispinis*
Brown cat shark—*Apristurus brunneus*
Brown smoothhound—*Mustelus henlei*
Bull shark—*Carcharhinus leucas*
Bullhead sharks—members of the family Heterodontidae
Carpet sharks—members of the order Orectolobiformes
Chain dogfish—*Scyliorhinus retifer*
Cigar shark (also Spined Pygmy shark)—*Squaliolus laticaudus*
Cookie-cutter shark—*Isistius brasiliensis*
Crocodile shark—*Pseudocarcharias kamoharai*
Cuban dogfish—*Squalus cubensis*
Dusky shark—*Carcharhinus obscurus*

False cat shark—*Pseudotriakis microdon*
Filetail cat shark—*Parmaturus xaniurus*
Finetooth shark—*Carcharhinus isodon*
Frill shark—*Chlamydoselachus anguineus*
Galapagos bullhead[2]—*Heterodontus quoyi*
Galapagos shark[2]—*Carcharhinus galapagensis*
Ganges shark[2]—*Glyphis gangeticus*
Goblin shark[2]—*Mitsukurina owstoni*
Gray reef shark[2]—*Carcharhinus amblyrhynchos*
Gray smoothhound—*Mustelus californicus*
Great hammerhead—*Sphyrna mokarran*
Green lanternshark[2]—*Etmopterus virens*
Greenland shark—*Somniosus microcephalus*
Horn shark—*Heterodontus francisci*
Houndsharks—sharks of the family Triakidae or genus *Mustelus*
Japanese sawshark—*Pristiophorus japonicus*
Kitefin shark—*Dalatias licha*
Largetooth cookie-cutter[2]—*Isistius plutodus*
Lemon shark—*Negaprion brevirostris*
Leopard shark—*Triakis semifasciata*
Longfin mako—*Isurus paucus*
Mackerel sharks—members of the family Lamnidae
Megamouth[2]—*Megachasma pelagios*
Mexican horn shark[2]—*Heterodontus mexicanus*
Narrowtooth shark—*Carcharhinus brachyurus*
Night shark—*Carcharhinus signatus*
Nurse shark—*Ginglymostoma cirratum*
Oceanic whitetip shark—*Carcharhinus longimanus*[3]
Pacific angel shark—*Squatina californica*
Pacific sharpnose shark—*Rhizoprionodon longurio*
Pacific sleeper shark—*Somniosus pacificus*
Pelagic thresher[2]—*Alopias pelagicus*
Porbeagle—*Lamna nasus*
Port Jackson shark[2]—*Heterodontus portusjacksoni*
Portuguese shark—*Centroscymnus coelolepis*
Prickly shark—*Echinorhinus cookei*
Pygmy ribbontail cat shark—*Eridacnis radcliffei*
Ragged-tooth shark—*Odontaspis ferox*
Reef shark—*Carcharhinus perezi*
Requiem sharks—members of the family Carcharhinidae
Salmon shark—*Lamna ditropis*
Sand tiger—*Eugomphodus taurus*
Sandbar shark—*Carcharhinus plumbeus*
Sawsharks—sharks of the family Pristiophoridae
Scalloped hammerhead—*Sphyrna lewini*
School shark (also Soupfin shark)—*Galeorhinus galeus*
Sevengill shark—*Notorynchus cepedianus*
Sharpnose sevengill shark[2]—*Heptranchias perlo*

Shortfin mako—*Isurus oxyrinchus*
Sicklefin lemon shark[2]—*Negaprion acutidens*
Sicklefin smoothhound—*Mustelus lunulatus*
Silky shark—*Carcharhinus falciformis*
Silvertip shark—*Carcharhinus albimarginatus*
Sixgill shark—*Hexanchus griseus*
Smalltail shark—*Carcharhinus porosus*
Smooth dogfish—*Mustelus canis*
Smooth hammerhead—*Sphyrna zygaena*
Soupfin shark (also School shark)—*Galeorhinus galeus*[2]
Spadenose shark—*Scoliodon laticaudus*
Spinner shark—*Carcharhinus brevipinna*
Spiny dogfish—*Squalus acanthias*
Spitting shark—*Nebrius ferrugineus*
Spotted wobbegong[2]—*Orectolobus maculatus*
Swell shark—*Cephaloscyllium ventriosum*
Thresher shark—*Alopias vulpinus*
Tiger shark—*Galeocerdo cuvier*
Whale shark—*Rhincodon typus*
Whaler sharks—members of the genus *Carcharhinus*
White shark—*Carcharodon carcharias*
Whitetip reef shark[2]—*Triaenodon obesus*
Winghead shark[2]—*Eusphyra blochii*
Wobbegongs[2]—members of the family Orectolobidae
Zebra shark[2]—*Stegostoma fasciatum*

1. Robins, C. Richard, chairman, et al. 1980. 4th edition. A List of Common and Scientific Names of Fishes from the United States and Canada. *American Fisheries Society Special Publication No. 12.*

2. Compagno, Leonard, J. V. 1984. Sharks of the World. *FAO Species Catalogue,* FAO Fisheries Synopsis No. 125, Vol. 4, Part 1 and 2. Rome: United Nations Development Programme, Food and Agriculture Organization of the United Nations.

3. The oldest scientific name for this species is *Carcharhinus maou.* Although correct, this name has rarely been used. We hope the name will be suppressed to avoid confusion.

Appendix 3

LENGTHS OF SELECTED SHARKS

unless otherwise indicated, lengths are total lengths (in meters; 1 meter = approximately 3.3 ft.)

* = actual measured length
? = questionable length

	Smallest free-living	At maturity		Maximum total length
		male	female	
Cigar Shark (*Squaliolus laticaudus*)	not reported	0.15	0.17–0.25	0.25*
Cookie-Cutter (*Isistius brasiliensis*)	not reported	0.31–0.37	0.38–0.44	0.44*
Spiny Dogfish (*Squalus acanthias*)	0.22–0.23	0.59–1.0	0.70–1.24	1.6*
Horn Shark (*Heterodontus francisci*)	0.15–0.16	0.58–0.71	>0.58	1.22*
Pacific Angel Shark (*Squatina californica*)	0.21–0.26	0.75–0.80	0.86–1.08	1.52*

	Smallest free-living	At maturity		Maximum total length
		male	female	
Spotted Wobbegong (*Orectolobus maculatus*)	0.21	0.60	not reported	3.2*
Sand Tiger (*Eugomphodus taurus*)	0.95–1.05	2.20–2.57	2.20–>3.0	3.18*
Lemon Shark (*Negaprion brevirostris*)	0.60–0.65	2.24	2.39	2.96*–3.4?
Zebra Shark (*Stegostoma fasciatum*)	0.20–0.36	1.47–1.83	1.69–1.71	3.0*–3.54?
Bull Shark (*Carcharhinus leucas*)	0.56–0.81	1.57–2.26	1.80–2.3	3.2*–3.4?
Porbeagle (*Lamna nasus*)	0.60–0.75	2.19–2.62	1.52–2.19	>3.0*–3.6?
Shortfin Mako (*Isurus oxyrinchus*)	0.60–0.70	1.95	2.80	3.28*–4.0?
Nurse Shark (*Ginglymostoma cirratum*)	0.27–0.30	2.25	2.30–2.40	3.04*–4.3
Thresher Shark (*Alopias vulpinus*)	1.10–1.5	3.19	3.76	5.70*–7.60?
Smooth Hammerhead (*Sphyrna zygaena*)	0.50–0.60	2.10–2.56	2.1–3.0	3.7?–4.0?
Great Hammerhead (*Sphyrna mokarran*)	0.50–0.70	2.34–2.69	2.50–3.0	5.49*–6.1
Tiger Shark (*Galeocerdo cuvier*)	0.70–0.76	2.26–3.10	3.15–3.20	5.9*–7.4? (9.1)?
Megamouth (*Megachasma pelagios*)		(only two individuals caught, both males)		4.5*
Blue Shark (*Prionace glauca*)	0.34–.53	1.8–2.8	2.2–3.2	3.8* (6.5?)
White Shark (*Carcharodon carcharias*)	about 1.0–1.10	3.0–4.0 fork length	about 4.7	6.0*–8.0?
Basking Shark (*Cetorhinus maximus*)	1.70	4.0–5.0	8.1–9.8	12.2–15.2?
Whale Shark (*Rhincodon typus*)	0.55–0.56	not reported		12.1* (18?)

Glossary

AMPHISTYLIC—condition in which jaws are suspended directly from the braincase.
AMPULLAE OF LORENZINI—skin pores in the head region containing receptor cells sensitive to electric fields.
ANAL FIN—a median fin on the lower surface of a shark positioned between the external opening of the cloaca and the base of the tail fin (absent in many species of sharks, particularly the Squaliformes).
ANNULI—growth rings formed in the calcified portions of some sharks; variously believed, depending on investigator and species of shark, to form at the rate of one or two per year.
ANTERIOR—situated in front of or toward the front (head).
CARTILAGE—gristle-like supporting connective tissue that, except for the teeth and denticles, forms the skeleton of sharks.
CAUDAL FIN—the tail fin.
CENTRUM—the main structural element of a vertebra, round in cross section.
CLASPER—rod-like modification of the pelvic fins of male sharks; used during mating.
CUSP—points on the surface of a tooth.
CLOACA—a collecting chamber for the digestive, urinary, and reproductive systems that opens externally between the pelvic fins.
DENTICLES—hard, tooth-like, placoid scales embedded in the skin.
DORSAL—situated near or on the back.
DORSAL FIN—the anteriormost fin on the back; there may be two: the anterior usually larger than the posterior, or second, dorsal fin.
ELASMOBRANCHOLOGISTS—(not in common use) scientists who study sharks, skates, rays, and chimaeras.

ELECTRIC FIELD—the distribution of the effects (charge force) of the electric charge on a body in the space surrounding the body.

EMBRYO—the developing young before birth or hatching.

FORK LENGTH—the length of the shark measured as a straight line from the tip of the snout to the posterior edge of the fork of the tail (usually measured as the distance between projections perpendicular to these two points).

GESTATION—period of development of embryo within the female.

HYOSTYLIC—condition in which the jaws are suspended from structures (hyomandibulae; one on each side), which are attached to the braincase (typical of most sharks).

ICHTHYOLOGISTS—scientists who study fishes.

KEEL—a ridge on the side of the body near the base of the caudal fin.

LATERAL—on or toward the side.

OOPHAGY (also OVIPHAGY)—a process within the uterus during which embryos derive their nourishment by eating other embryos and unfertilized eggs.

OVIPAROUS—a mode of reproduction in which fertilized eggs are laid and hatched outside the female's body.

OVOVIVIPAROUS—a mode of reproduction in which the young hatch from eggs and develop within the female's body (a placental connection with the mother is absent).

PECTORAL FINS—the anteriormost pair of fins, located on the sides of the body behind and/or below the gill slits.

PELAGIC—existing in the open ocean.

PELVIC FINS—the pair of fins on the undersurface of the shark body behind the pectoral fins and lying to either side of the external opening of the cloaca.

SPIRACLE—opening behind the eye (remnant of a gill slit present in ancient elasmobranchs).

TAPETUM LUCIDUM—structure situated behind the retina of the eye, consisting of reflecting plates and associated mobile cells; used to enhance vision.

TOTAL LENGTH—the length of a shark measured from the tip of the snout along an imaginary line through the long axis of the body, to a point on a line perpendicular to the tip of the upper lobe of the caudal fin.

VENTRAL—toward or on the undersurface of the body.

VERTEBRATE—animal with a backbone.

VIVIPAROUS—a mode of reproduction in which the young hatch, develop, and are nourished within the female by a placental attachment.

Bibliography

General and Popular

Bigelow, H. B., and W. C. Schroeder

1948. Sharks. *In Fishes of the Western North Atlantic, Memoir of the Sears Foundation for Marine Research.* New Haven: Yale University. No. 1, part 1, pp. 59–576.

1953. Sawfishes, guitarfishes, skates and rays. *In Fishes of the Western North Atlantic, Memoir of the Sears Foundation for Marine Research.* New Haven: Yale University. No. 1, part 2, pp. 1–514. Chimaeroids, ibid., pp. 515–562.

Budker, P.

1971. *The Life of Sharks.* London: Weidenfeld and Nicolson, 222 pp.

Cartmell, B. C.

1978. *Let's Go Fossil Shark Tooth Hunting.* Ann Arbor, Michigan: Natural Science Research, 76 pp.

Casey, J. G.

1964. Anglers' Guide to Sharks of the Northeastern United States, Maine to Chesapeake Bay. *U.S. Fish and Wildlife Service, Bureau of Sport Fisheries and Wildlife, Circular* 179, 32 pp.

Castro, J. I.

1983. *The Sharks of North American Waters.* College Station: Texas A&M University Press, 180 pp.

Clark, E.

1969. *The Lady and the Sharks.* New York: Harper & Row, 269 pp.

1981. Sharks Magnificent and Misunderstood. *National Geographic,* 160(2): 13–83.

Compagno, L. J. V.
1984. Sharks of the World. FAO Species Catalogue. *FAO Fisheries Synopsis* No. 125, vol. 4, part 1 and part 2. Rome: United Nations Development Programme, Food and Agriculture Organization of the United Nations, 655 pp.

Daniel, J. F.
1934. *The Elasmobranch Fishes.* Berkeley: University of California Press, 332 pp. (Many later editions published.)

Dingerkus, G.
1985. *The Shark Watcher's Guide.* New York: Wanderer Books, Simon & Schuster, Inc., 144 pp.

Ellis, R.
1975. *The Book of Sharks.* New York: Grosset & Dunlap, 256 pp.

Ellis, R., and J. McCosker
1986. Speaking of sharks. *Oceans,* 19(3): 25–29, 58–60.

Eschmeyer, W. N., E. S. Herald, and H. Hammann
1983. *A Field Guide to Pacific Coast Fishes of North America.* Boston: Houghton Mifflin Co., 336 pp.

Gilbert, P. W., ed.
1963. *Sharks and Survival.* Boston: D.C. Heath and Co., 578 pp.

Gilbert, P. W., R. F. Mathewson, and D. P. Rall, eds.
1967. *Sharks, Skates and Rays.* Baltimore: Johns Hopkins University Press, 624 pp.

Grant, E. M.
1982. *Guide to Fishes.* Brisbane, Queensland: Department of Harbours and Marine, 896 pp.

Hoar, W. S., and D. J. Randall, eds.
1969–1978. *Fish Physiology.* Vol. I, 1969, *Excretion, Ionic Regulation, and Metabolism,* 465 pp.; Vol. II, 1969, *The Endocrine System,* 446 pp.; Vol. III, 1969, *Reproduction and Growth, Bioluminiscense, Pigments and Poisons,* 485 pp.; Vol. IV, 1970, *The Nervous System, Circulation and Respiration,* 532 pp.; Vol. V, 1971, *Sensory Systems and Electric Organs,* 600 pp.; Vol. VI, 1971, *Environmental Relations and Behavior,* 559 pp.; Vol. VII, 1978, *Locomotion,* 576 pp. New York: Academic Press.

Hoar, W. S., D. J. Randall, and J. R. Brell, eds.
1978. *Fish Physiology.* Vol. VIII, *Bioenergetics and Growth.* New York: Academic Press, 786 pp.

Hoar, W. S., D. J. Randall, and E. M. Donaldson, eds.
1983. *Fish Physiology.* Vol. IX, *Reproduction,* Part A, *Endocrine Tissues and Hormones,* 483 pp.; Vol. IX, *Reproduction,* Part B, *Behavior and Fertility Control.* New York: Academic Press, 476 pp.

Hodgson, E. S., and R. F. Mathewson, eds.
1978. *Sensory Biology of Sharks, Skates, and Rays.* Arlington, Virginia: Office of Naval Research, 666 pp.

Johnson, R. H.
1978, 1984. *Sharks of Tropical and Temperate Seas.* Papeete, Tahiti: Les Editions du Pacifique, 170 pp.

Lineaweaver, T. H., III, and R. H. Backus
1970. *The Natural History of Sharks.* New York: J.B. Lippincott Co., 256 pp. (1984 paperback edition, Schocken Books).

Macdonald, J. R.
1983. *The Fossil Collector's Handbook: A Paleontology Field Guide.* Englewood Cliffs, New Jersey: Prentice Hall, 193 pp.

McCormick, H. W., T. Allen, and W. Young
1963. *Shadows in the Sea: Sharks, Skates, and Rays.* Baltimore: Johns Hopkins University Press, 415 pp.

Moss, S. A.
1984. *Sharks—An Introduction for the Amateur Naturalist.* Englewood Cliffs, New Jersey: Prentice-Hall, Inc., 246 pp.

Northcutt, R. G., ed.
1977. Recent Advances in the Biology of Sharks. *American Zoologist,* 17(2): 287–515.

Prince, E. D., and L. M. Pulos, eds.
1983. Proceedings of the International Workshop on Age Determination of Oceanic Pelagic Fishes: Tunas, Billfishes, and Sharks. *U.S. Dept. of Commerce, NOAA Technical Report NMFS 8,* 211 pp.

Randall, J. E.
1986. *Sharks of Arabia.* London: Immel Publishing, 148 pp.

Reader's Digest
1986. *Sharks, Silent Hunters of the Deep.* Sydney, London, New York, Montreal, Capetown: Reader's Digest Services Pty Ltd, 208 pp.

Romer, A. S.
1966. *Paleontology.* 3d edition. Chicago and London: University of Chicago Press, 468 pp.

Schwartz, F. J.
1984. *Sharks, Sawfish, Skates, and Rays of the Carolinas.* Special Publication, Institute of Marine Sciences, Moorehead City, North Carolina, 101 pp.

Seaman, W., ed.
1976. Sharks and Man: A Perspective. *Florida Sea Grant Program, Report No.* 10. Gainesville, Florida, 36 pp.

Sibley, G., J. A. Seigel, and C. C. Swift, eds.
1985. Biology of the White Shark. *Memoirs of the Southern California Academy of Sciences,* vol. 9, 155 pp.

Stead, D. G.
1964. *Sharks and Rays of Australian Seas.* London: Angus and Robertson, 211 pp.

Steel, R.
1985. *Sharks of the World.* New York: Facts on File Publications, 192 pp.

Stevens, J. D., ed.
1987. *Sharks.* New York: Facts on File Publications, 240 pp.

Thompson, I.
1982. *The Audubon Society Field Guide to North American Fossils.* New York: Alfred A. Knopf, 846 pp.

Thompson, J. R., and S. Springer
1965. Sharks, Skates, Rays, and Chimaeras. *U.S. Fish and Wildlife Service, Circ. 228,* 19 pp.

Thorson, T. B., ed.
1976. *Investigations of the Ichthyofauna of Nicaraguan Lakes.* University of Nebraska, Lincoln: School of Biological Sciences, 663 pp.

Wilson, R., and J. Q. Wilson
1985. *Watching Fishes: Life and Behavior on Coral Reefs.* New York: Harper & Row, 275 pp.

Woods Hole Oceanographic Institution
1981/82. Sharks. *Oceanus,* 24(4): 80

Zahuranec, B. J., ed.
1983. Shark Repellents from the Sea: New Perspectives. *American Association for the Advancement of Science, Selected Symposium* 83, 210 pp.

Further Readings

Aasen, O.
1963. Length and Growth of the Porbeagle (*Lamna nasus,* Bonnaterre) in the North West Atlantic. *Fiskeridirektoratets Skrifter,* 13(6): 20–37.

Ainley, D. G., R. P. Henderson, H. R. Huber, R. J. Boekelheide, S. G. Allen, and R. L. McElroy
1985. Dynamics of White Shark/Pinniped Interactions in the Gulf of the Farallones. *Memoirs of the Southern California Academy of Sciences,* vol. 9, pp. 109–122.

Al-Badri, M., and R. Lawson
1985. Contribution on the Taxonomy of the Spiny Dogfish *Squalus acanthias* L. *Cybium,* 9(4): 385–399.

Anderson, E. D.
1985. Analysis of Various Sources of Pelagic Shark Catches in the Northwest and Western Central Atlantic Ocean and Gulf of Mexico with Comments on Catches of Other Large Pelagics. *U.S. Dept. Commerce, NOAA Technical Report NMFS* 31, pp. 1–15.

Andrews, S. M., B. G. Gardiner, R. S. Miles, and C. Patterson, contributors
1967. Chapter 26, Pisces *In* Geological Society of London, *The Fossil Record, a Symposium with Documentation,* pp. 637–683.

Baldridge, H. D.
1972. Accumulation and Function of Liver Oil in Florida Sharks. *Copeia,* no. 2, pp. 306–325.

1973. *Shark Attack against Man, a Program of Data Reduction and Analysis.* Sarasota, Florida: Mote Marine Laboratory, 66 pp.

1974. *Shark Attack.* New York: Berkeley Pub. Corp., 263 pp.

Bass, A. J.
1977. Long-term Recoveries of Tagged Sharks. *Copeia,* no. 3, pp. 574–575.

Bass, A. J., J. D. D'Aubrey, and N. Kistnasamy
1975a. Sharks of the East Coast of Southern Africa, 3: The Families Carcharhinidae (excluding *Mustelus* and *Carcharhinus*) and Sphyrnidae.

Investigational Report 38, 100 pp. Durban: Oceanographic Research Institute.

1975b. Sharks of the East Coast of Southern Africa, 4: The Families Odontaspididae, Scapanorhynchidae, Isuridae, Cetorhinidae, Alopiidae, Orectolobidae and Rhiniodontidae. *Investigational Report* 39, 102 pp. Durban: Oceanographic Research Institute.

1975c. Sharks of the East Coast of Southern Africa, 5: The Families Hexanchidae, Chlamydoselachidae, Heterodontidae, Pristiophoridae and Squatinidae. *Investigational Report* 43, 50 pp. Durban: Oceanographic Research Institute.

Baughman, J. L.
1955. The Oviparity of the Whale Shark, *Rhincodon typus,* with Records of This and Other Fishes in Texas Water. *Copeia,* no. 1, pp. 54–55 and plate I.

Bedford, D. W.
1987. Sexual Maturity and Fecundity in the Common Thresher Shark (*Alopias vulpinus*) from Southern California Waters. *American Society of Ichthyologists and Herpetologists, Program and Abstracts, 67th Annual Meeting,* 21–26 June 1987, Albany, New York, pp. 32–33.

Benz, G. W.
1985. Copepods as Parasites of Sharks [Abstract]. *In Sharks: Recent Advances in Captive Biology, a Symposium on the Captive Maintenance of Elasmobranchs.* The National Aquarium in Baltimore and Department of Pathology, University of Maryland School of Medicine.

Blake, R. W.
1983. *Fish Locomotion.* Cambridge, London: Cambridge University Press, 208 pp.

Boord, R. L., and C. B. G. Campbell
1977. Structural and Functional Organization of the Lateral Line System of Sharks. *American Zoologist,* 17(2): 431–440.

Branstetter, S., J. A. Musick, and J. A. Colvocoresses
1987. A Comparison of the Age and Growth of the Tiger Shark, *Galeocerdo cuvieri* from off Virginia and from the northwestern Gulf of Mexico. *Fishery Bulletin,* 85(2): 269–279.

Branstetter, S., and R. Stiles
1987. Age and Growth Estimates of the Bull Shark, *Carcharhinus leucas,* from the Northern Gulf of Mexico. *Environmental Biology of Fishes,* 20(3): 169–181.

Brownlee, S.
1985. Special Report: On the Track of the Real Shark. *Discover,* 6(7): 26–38.

Cadenat, J., and J. Blache
1981. Requins de Méditerranée et d'Atlantique. *Faune Tropical No. XXI.* Paris: Office de la Recherche Scientifique et Technique Outre Mer (ORSTOM), 330 pp.

Cailliet, G. M., L. J. Natanson, B. A. Welden, and D. A. Ebert
1983. Preliminary Studies on the Age and Growth of the Blue, *Prionace glauca,* Common Thresher, *Alopias vulpinus,* and Shortfin Mako, *Isurus oxyrinchus,* Sharks from California Waters. *U.S Dept. of Commerce, NOAA Technical Report NMFS* 8, pp. 179–188.

1985. Preliminary Studies on the Age and Growth of the White Shark, *Carcharodon carcharias*, Using Vertebral Bands. *Memoirs of the Southern California Academy of Sciences*, vol. 9, 24 May, pp. 49–60.

Cailliet, G. M., L. K. Martin, D. Kusher, P. Wolf, and B. A. Welden
1983. Techniques for Enhancing Vertebral Bands in Age Estimation of California's Elasmobranchs. *U.S. Dept. of Commerce, NOAA Technical Report NMFS 8*, pp. 157–165.

Caplan, A. J.
1984. Cartilage. *Scientific American*, 251(4): 84–94.

Cappetta, H.
1987. Chondrichthyes II. Mesozoic and Cenozoic Elasmobranchii. *Handbook of Paleoichthyology*. Vol. 3B. Stuttgart, New York: Gustav Fischer Verlag, 193 pp.

Carey, F. G., and J. Scharold
1987. Tracking Blue Sharks in Course and Depth. *American Society of Ichthyologists and Herpetologists, Program and Abstracts, 67th Annual Meeting*, 21–26 June, Albany, New York, p. 36.

Carey, F. G., and J. M. Teal
1969. Mako and Porbeagle: Warm Bodied Sharks. *Comparative Biochemical Physiology*, vol. 28, pp. 199–204.

Carey, F. G., J. M. Teal, and J. W. Kanwisher
1981. The Visceral Temperature of Mackerel Sharks (Lamnidae). *Physiological Zoology*, 54(3): 334–344.

Carey, F. G., J. W. Kanwisher, O. Brazier, G. Gabrielson, J. G. Casey, and H. S. Pratt, Jr.
1982. Temperature and Activities of a White Shark, *Carcharodon carcharias*. *Copeia*, no. 2, pp. 254–260.

Carey, F. G., J. G. Casey, H. L. Pratt, D. Urquhart, and J. E. McCosker
1985. Temperature, Heat Production and Heat Exchange in Lamnid Sharks. *Memoirs of the Southern California Academy of Sciences*, vol. 9, 24 May, pp. 92–108.

Carrier, J. C.
1987. Contributions to the Biology of the Nurse Shark (*Ginglymostoma cirratum* Bonnaterre): Growth, Movement and Ectoparasites. *American Society of Ichthyologists and Herpetologists, Program and Abstracts, 67th Annual Meeting*, 21–26 June, Albany, New York, p. 37.

Casey, J. G., H. W. Pratt, and C. Stillwell
1980–1983. The Shark Tagger. *Newsletter of the Cooperative Shark Tagging Program*, (1)1980, (2)1981, (3)1982, (4)1983. Narragansett, Rhode Island: National Marine Fisheries Service.

Casey, J. G., H. W. Pratt, C. Stillwell, and N. Kohler
1984–1987. The Shark Tagger. *Newsletter of the Cooperative Shark Tagging Program*, (5)1984, (6)1985, (7)1986, (8)1987. Narragansett, Rhode Island: National Marine Fisheries Service.

Casey, J. G., H. L. Pratt, Jr., and C. E. Stillwell
1983. Age and Growth of the Sandbar Shark, *Carcharhinus plumbeus*, from the Western North Atlantic. *U.S. Dept. of Commerce, NOAA Technical Report NMFS 8*, pp. 189–191.

Casey, J. G., and C. E. Stillwell

1983. Apex Predators: A Growing Fishery and a Growing Need for Biological Information for the Wise Management of the Northwest Atlantic's Large Sharks, Tunas, Billfishes, and Swordfish. [Mimeo.] Narragansett, Rhode Island: Apex Predators Investigation, Marine Ecosystems Division, National Marine Fisheries Service, 11 pp.

Casey, J. G., and J. J. Hoey

1985. Estimated Catches of Large Sharks by U.S. Recreational Fishermen in the Atlantic and Gulf of Mexico. *U.S. Dept. Commerce, NOAA Technical Report NMFS* 31, pp. 15–19.

Casey, J. G., and H. L. Pratt, Jr.

1985. Distribution of the White Shark, *Carcharodon carcharias,* in the Western North Atlantic. *Memoirs of the Southern California Academy of Sciences,* vol. 9, 24 May, pp. 2–14.

Castro, J. I.

1987. The Biology of the Smalleye Hammerhead (*Sphyrna tudes*) off Trinidad. *American Society of Ichthyologists and Herpetologists, Program and Abstracts, 67th Annual Meeting,* 21–26 June, Albany, New York, p. 37.

Clark, E.

1947. Notes on the Inflating Power of the Swell Shark *Cephaloscyllium uter. Copeia,* no. 4, pp. 278–280.

1959. Instrumental Conditioning of Lemon Sharks. *Science,* 130(3368): 217–218.

1963. Maintenance of Sharks in Captivity with a Report on Their Instrumental Conditioning. *In* P. W. Gilbert, ed., *Sharks and Survival.* Boston: D.C. Heath Co., pp. 115–149.

1975. Into the Lairs of "Sleeping Sharks." *National Geographic,* 47(4): 570–584.

Clark, E., and K. v. Schmitt

1965. Sharks of the Central Gulf Coast of Florida. *Bulletin of Marine Science,* 15(1): 13–83.

Cohen, J. L.

1981. Vision in Sharks. *Oceanus,* 24(4): 17–22.

Compagno, L. J. V.

1973. Interrelationships of Living Elasmobranchs. *In* P. H. Greenwood, R. S. Miles, and C. Patterson, eds., Interrelationships of fishes. *Zoological Journal Linnean Society of London,* vol. 53, Suppl. 1., pp. 15–61.

1977. Phyletic Relationships of Living Sharks and Rays. *American Zoologist,* 17(2): 303–322.

1981. Legend versus Reality: The Jaws Image and Shark Diversity. *Oceanus,* 24(4): 5–16.

Coppleson, V. M.

1962. *Shark Attack.* 2d ed. Sydney: Angus and Robertson, 269 pp.

Darom, D., and A. Baranes

1980. The Shark [In Hebrew]. Israel: Massada Ltd., 119 pp.

D'Aubrey, J. D.

1964. A Carchariid Shark New to South African Waters. *Investigational Report*

No. 9. Durban: Oceanographic Research Institute, 16 pp.

Davies, D. H.
1964. *About Sharks and Shark Attack.* Pietermaritzburg, Natal: Shuter & Shooter, 237 pp.

Davis, C.
1983. The Awesome Basking Shark. *Sea Frontiers,* 29(2): 78–85.

Dempster, R. P., and E. S. Herald
1961. Notes on the Hornshark, *Heterodontus francisci,* with Observations on Mating Activities. *Occasional Papers of the California Academy of Sciences,* no. 33, pp. 1–7.

Diamond, J. M.
1985. Filter-feeding on a Grand Scale. *Nature,* vol. 316, 22 August, pp. 679–680.

Dodd, J. M.
1983. Reproduction in Cartilaginous Fishes (Chondrichthyes). *In* W. S. Hoar, D. J. Randall, and E. M. Donaldson, eds., *Fish Physiology,* Vol. IX, part A. New York: Academic Press, pp. 31–95.

Dor, M.
1984. *Checklist of the Fishes of the Red Sea.* Jerusalem: Israeli Academy of Sciences and Humanities, 437 pp.

Eibl-Eibesfeldt, I., and H. Hass
1959. Erfahrungen mit Haien. *Zeitschrift für Tierpsychologie,* 16(6): 733–746.

Garrick, J. A. F.
1964. Additional Information on the Morphology of an Embryo Whale Shark. *Proceedings of the United States National Museum;* 115(3476): 1–8.

1982. Sharks of the Genus *Carcharhinus. U.S. Dept. of Commerce, NOAA Technical Report* NMFS Circ. 445, 194 pp.

1985. Additions to a Revision of the Shark Genus *Carcharhinus:* Synonymy of *Aprionodon* and *Hypoprion,* and Description of a New Species of *Carcharhinus. U.S. Dept. of Commerce; NOAA Technical Report NMFS* 34, 26 pp.

Garrick, J. A. F., and S. Springer
1964. *Isistius plutodus,* a New Squaloid Shark from the Gulf of Mexico. *Copeia,* no. 4, pp. 678–682.

Geologic Names Committee
1984. Geologic Time Chart. Stratigraphic Notes, 1983, *United States Geological Survey Bulletin,* 1537–A, pp. A1–A4.

Gilbert, C. R.
1967. A Revision of the Hammerhead Sharks (Family Sphyrnidae). *Proceedings of the United States National Museum,* 119(3539): 1–77.

Gilbert, P. W.
1963. The Visual Apparatus of Sharks. *In* P. W. Gilbert, ed., *Sharks and Survival.* Baltimore: Johns Hopkins University Press, pp. 283–321.

1977. Two Decades of Shark Research: A Review. *Bioscience,* 27(10): 670–673.

1981–82. Patterns of Shark Reproduction. *Oceanus,* 24(4): 30–39.

1984a. Les Requins. *La Recherche*, 15(157): 925–936. (in French)

1984b. Biology and Behavior of Sharks. *Endeavor*, new series, 8(4): 179–187.

Gilbert, P. W., J. G. Swak, and R. E. Pelham
1981. Rapid Pupil Change in Selachians. *Journal of Zoology*, 59(3): 560–564.

Gilmore, R. G.
1983. Observations on the Embryos of the Longfin Mako, *Isurus paucus*, and the Bigeye Thresher, *Alopias superciliosus*. *Copeia*, no. 2, pp. 375–382.

Gilmore, R. G., J. W. Dodrill, and P. A. Linley
1983. Reproduction and Embryonic Development of the Sandtiger Shark, *Odontaspis taurus* (Rafinesque). *Fishery Bulletin*, NOAA/NMFS, 81(2): 201–225.

Gordievskaya, V. S.
1973. Shark Flesh in the Food Industry. Pacific Scientific Research Institute of Marine Fisheries and Oceanography [translated from Russian], published for the National Marine Fisheries Service, NOAA, by the Israel Program for Scientific Translations, Jerusalem, 26 pp.

Graeber, R. C.
1978. Behavioral Studies Correlated with Central Nervous System Integration of Vision in Sharks. *In* E. S. Hodgson and R. W. Mathewson, eds., *Sensory Biology of Sharks, Skates, and Rays*. Arlington, Virginia: Office of Naval Research, pp. 195–225

Grimes, D. J., P. Brayton, B. Youngren-Grimes, M. T. MacDonnell, M. V. Parsey, S. H. Gruber, and R. R. Colwell
1985. The Bacterial Flora of Sharks. *American Society of Ichthyologists and Herpetologists, 65th Annual Meeting, Program and Abstracts*. 9–14 June, University of Tennessee, Knoxville.

Gruber, S. H.
1967. A Behavioral Measurement of Dark Adaptation in the Lemon Shark, *Negaprion brevirostris*. *In* P. W. Gilbert, R. F. Mathewson, and D. P. Ralls, eds. *Sharks, Skates, and Rays*. Baltimore: Johns Hopkins University Press, pp. 479–490.

1977. The Visual System of Sharks: Adaptations and Capability. *American Zoologist*, 17(2): 453–469.

1981a. Shark Repellents: Perspectives for the Future. *Oceanus*, 24(4): 72–76.

1981b. Lemon Sharks: Supply-side Economists of the Sea. *Oceanus*, 24(4): 56–64.

1985. Bioenergetics of the Lemon Shark, *Negaprion brevirostris* (Poey) under Laboratory and Field Conditions. *Abstracts of Papers, Second International Conference on Indo-Pacific Fishes*, Tokyo, July 28–August 3, p. 52.

1988. Sharks of the Shallows. *Natural History*, 97(3): 50–57.

Gruber, S. H., and A. A. Myrberg, Jr.
1977. Approaches to the Study of the Behavior of Sharks. *American Zoologist*, 17(2): 471–486.

Gruber, S. H., and J. L. Cohen
1978. Visual System of the Elasmobranchs: State of the Art 1960–1975. *In* E. S. Hodgson and R. W. Mathewson, eds., *Sensory Biology of Sharks, Skates, and Rays*. Arlington, Virginia: Office of Naval Research, pp. 11–105.

Gruber, S. H., and P. Morrissey
1978. Sharks: Good Vision or Poor? *Sea Frontiers*, 24(4): 229–236.

Gruber, S. H., and R. G. Stout
1983. Biological Materials for the Study of Age and Growth in a Tropical Marine Elasmobranch, the Lemon Shark, *Negaprion brevirostris* (Poey). *U.S. Dept. of Commerce, NOAA Technical Report NMFS 8*, pp. 193–205.

Gubanov, Y. P.
1972. On the Biology of the Thresher Shark (*Alopias vulpinus* (Bonnaterre)) in the Northwest Indian Ocean. *Journal of Ichthyology* [Engl. transl. from Voprosy Ikhtiologii], 12(4): 591–600.

Gudger, E. W.
1915. Natural History of the Whale Shark, *Rhincodon typus* Smith. *Zoologica*, 1(19): 349–389.

1941. The Food and Feeding Habits of the Whale Shark *Rhincodon typus*. *Journal of the Elisha Mitchell Scientific Society*, vol. 57, pp. 57–72.

1949. Natural History Notes on Tiger Sharks, *Galeocerdo tigrinus*, Caught at Key West, Florida, with Emphasis on Food and Feeding Habits. *Copeia*, no. 1, pp. 39–47.

Harding, J.
1969. Monster Shark. *Sea Frontiers*, 15(1): 50–53.

Heald, D.
1982. The WA Shark Fishery. *Fins*, 15(3): 16–21.

Heemstra, P. C.
1980. How Sharks Feed. *Sea Frontiers*, 26(4): 212–216.

Hodgson, E. S., and R. F. Mathewson
1978. Electrophysiological Studies of Chemoreception in Elasmobranchs. *In* E. S. Hodgson and R. F. Mathewson, eds., *Sensory Biology of Sharks, Skates, and Rays*. Arlington, Virginia: Office of Naval Research, pp. 227–267.

Hodgson, L.
1984–85. Shark Netting: A Costly Security Blanket. *Australian Natural History*, 21(7): 283–284.

Jensen, N. H.
1976. Reproduction of the Bull Shark, *Carcharhinus leucas*, in the Lake Nicaragua-Rio San Juan System. *In* T. Thorson, ed., *Investigations of the Ichthofauna of Nicaraguan Lakes*. University of Nebraska, Lincoln: School of Biological Sciences, pp. 539–559.

Johnson, C. S., B. L. Scronce, and M. W. McManus
1984. Detection of DC Electric Dipoles in Background Fields by the Nurse Shark. *Journal of Comparative Physiology A*, vol. 155, pp. 681–687.

Johnson, R. H., and D. R. Nelson
1973. Agonistic Display in the Gray Reef Shark, *Carcharhinus menisorrah*, and its Relationship to Attacks on Man. *Copeia*, no. 1, pp. 76–84.

1978. Copulation and Possible Olfaction-mediated Pair Formation in Two Species of Carcharhinid Sharks. *Copeia*, no. 3, pp. 539–542.

Jones, E. C.

1971. *Isistius brasiliensis*, a Squaloid Shark, the Probable Cause of Crater Wounds on Fishes and Cetaceans. *Fishery Bulletin*, 69(4): 791–798.

Kalmijn, A. J.

1971. The Electric Sense of Sharks and Rays. *Journal of Experimental Biology*, 55(2): 371–383.

1977. The Electric and Magnetic Sense of Sharks, Skates, and Rays. *Oceanus*, 20(3): 45–52.

1978. Electric and Magnetic Sensory World of Sharks, Skates, and Rays. *In* E. S. Hodgson and R. W. Mathewson, eds., *Sensory Biology of Sharks, Skates, and Rays*. Arlington, Virginia: Office of Naval Research, pp. 507–528.

1982. Electric and Magnetic Field Detection in Elasmobranch Fishes. *Science*, 218(4575): 916–918.

Kauffman, D. E.

1950. Notes on the Biology of the Tiger Shark (*Galeocerdo arcticus*) from Philippine Waters. *Research Report 16, U.S. Fish and Wildlife Service*, 10 pp.

Ketchen, K. S.

1986. The Spiny Dogfish (*Squalus acanthias*) in the Northeast Pacific and a History of its Utilization. *Canadian Special Publication of Fisheries and Aquatic Sciences* 88, 78 pp.

King, J. E.

1982. Fossils. *Story of Illinois Series, no. 14*. Springfield, Illinois: Illinois State Museum, 68 pp.

Klausewitz, W., and Hjalmar Thiel

1982. Tiefenwasser-und tiefseefische aus dem Roten Meer. VI. Über das vorkommen des haifisches *Iago omanensis* (NORMAN) (Pisces: Chondrichthyes: Elasmobranchii: Carcharinidae) und des messerzahnaals *Muraenesox cinereus* (FORSSKAL) (Teleostei: Apodes: Muraenesocidae), beide mit hilfe der fotofalle beobachtet und gefangen). *Senckenbergiana maritima*, 14(5/6): 227–243.

Klimley, A. P.

1980. Observations of Courtship and Copulation in the Nurse Shark, *Ginglymostoma cirratum*. *Copeia*, no. 4, pp. 878–882.

1981. Grouping Behavior in the Scalloped Hammerhead. *Oceanus*, 24(4): 65–71.

1985. The Areal Distribution and Autoecology of the White Shark, *Carcharodon carcharias*, off the West Coast of North America. *Memoirs of the Southern California Academy of Sciences*, vol. 9, 24 May, pp. 15–40.

Klimley, A. P., and D. R. Nelson

1981. Schooling of Hammerhead Sharks, *Sphyrna lewini*, in the Gulf of California. *Fishery Bulletin, NOAA/NMFS*, 79(2): 356–360.

Lavenberg, R. J., and J. A. Seigel

1985. The Pacific's Megamystery—Megamouth. *Terra*, 23(4): 29–31.

Lea, R. N., and D. J. Miller

1985. Shark Attacks off the California and Oregon Coasts: An Update, 1980–84. *Memoirs of the Southern California Academy of Sciences*, vol. 9, 24 May, pp. 136–149.

Le Boeuf, B. J., J. E. McCosker, and J. Hewitt

1987. Crater Wounds on Northern Elephant Seals: The Cookiecutter Shark Strikes Again. *Fishery Bulletin*, 85(2): 387–392.

Lee, A., and R. Langer

1983. Shark Cartilage Contains Inhibitors of Tumor Angiogenesis. *Science*, vol. 221, 16 September, pp. 1185–1187.

Maisey, J. G.

1984a. Chondrichthyan Phylogeny: A Look at the Evidence. *Journal of Vertebrate Paleontology*, 4(3): 359–371.

1984b. Higher Elasmobranch Phylogeny and Biostratigraphy. *Zoological Journal of the Linnean Society*, 82(1 and 2): 33–54.

1985. Cranial Morphology of the Fossil Elasmobranch *Synechodus dubrisiensis*. *American Museum Novitates*, no. 2804, January 30, pp. 1–28.

McCosker, J. E.

1981. Great White Shark. *Science 81*, vol. 2, July/August, pp. 42–51.

1985. White Shark Attack Behavior: Observations of and Speculations about Predator and Prey Strategies. *Memoirs of the Southern California Academy of Sciences*, vol. 9, 24 May, pp. 123–135.

1987. The White Shark, *Carcharodon carcharias*, Has a Warm Stomach. *Copeia*, no. 1, pp. 195–197.

McKibben, J. N., and D. R. Nelson

1986. Patterns of Movement and Grouping of Gray Reef Sharks, *Carcharhinus amblyrhynchos*, at Enewetak, Marshall Islands. *Bulletin of Marine Science*, 38(1): 89–110.

McManus, M. E., C. S. Johnson, and M. M. Jeffries

1984. Training Nurse Sharks Using Operant Conditioning. *Naval Ocean Systems Center Technical Report* 977, 12 pp.

Merrick, J. R., and G. E. Schmida

1984. *Australian Freshwater Fishes: Biology and Management.* North Ryde, N.S.W., Australia: J.R. Merrick, 409 pp.

Michael, S.

1986. The Inflatable Sharks. Survival of the Fattest? *Sea Frontiers*, 32(1): 27–33.

Miller, D. I., and R. S. Collier

1980. Shark Attacks in California and Oregon 1926–1979. *California Fish and Game*, vol. 67, pp. 76–104.

Morrissey, J. F.

1987. Shark Research at Sea. *Sea Frontiers/Sea Secrets*, 33(4): 244–255.

Moss, M. L.

1977. Skeletal Tissues in Sharks. *American Zoologist*, 17(2): 335–342.

Moss, S. A.

1967. Tooth Replacement in the Lemon Shark *Negaprion brevirostris*. *In*

P. W. Gilbert, R. F. Mathewson, and D. P. Rall, eds., *Sharks, Skates and Rays.* Baltimore: Johns Hopkins University Press, pp. 319–329.

1977. Feeding Mechanisms in Sharks. *American Zoologist,* 17(2): 355–364.

1981. Shark Feeding Mechanisms. *Oceanus,* 24(4): 23–29.

Moy-Thomas, J. A.
1971. *Palaeozoic Fishes.* 2d ed., revised by R. S. Miles. London: Chapman and Hall Ltd., 259 pp.

Myers, G.
1952. Sharks and Sawfishes in the Amazon. *Copeia,* no. 4, pp. 268–269.

Myrberg, A. A., Jr.
1978. Underwater Sound—Its Effect on the Behavior of Sharks. *In* E. S. Hodgson and R. W. Mathewson, eds., *Sensory Biology of Sharks, Skates, and Rays.* Arlington, Virginia: Office of Naval Research, pp. 391–417.

Myrberg, A. A., Jr., and S. H. Gruber
1974. The Behavior of the Bonnethead Shark, *Sphyrna tiburo. Copeia,* no. 2, pp. 358–374.

Myrberg, A. A. Jr., C. R. Gordon, and A. P. Klimley
1976. Attraction of Free Ranging Sharks to Low Frequency Sound, with Comments on its Biological Significance. *In* A. Schuijf and A. D. Hawkins, eds., *Sound Reception in Fishes.* Amsterdam, New York: Elsevier Scientific Pub., pp. 205–228.

Nammack, M. F., J. A. Musick, and J. A. Colvocoresses
1985. Life History of the Spiny Dogfish off the Northeastern United States. *Transactions of the American Fisheries Society,* vol. 114, pp. 367–376.

Natanson, L. J., and G. M. Cailliet
1986. Reproduction and Development of the Pacific Angel Shark. *Copeia,* no. 4, pp. 987–994.

Nelson, D. R.
1977. On the Field Study of Shark Behavior. *American Zoologist,* vol. 17, pp. 501–507.

1978. Telemetering Techniques for the Study of Free-ranging Sharks. *In* E. S. Hodgson and R. W. Mathewson, eds., *Sensory Biology of Sharks, Skates, and Rays.* Arlington, Virginia: Office of Naval Research, pp. 419–482.

1981. Aggression in Sharks: Is the Gray Reef Shark Different? *Oceanus,* 24(4): 45–55.

Nelson, D. R., and R. H. Johnson
1980. Behavior of the Reef Sharks of Rangiroa, French Polynesia. *National Geographic Society Research Reports,* vol. 12, pp. 479–499.

Nelson, D. R., R. R. Johnson, J. N. McKibben, and G. G. Pittenger
1986. Agonistic Attacks on Divers and Submersibles by Gray Reef Sharks, *Carcharhinus amblyrhynchos:* Antipredatory or Competitive? *Bulletin of Marine Science,* 38(1): 68–88.

Northcutt, R. G.
1977. Elasmobranch Central Nervous System Organization and its Possible Evolutionary Significance. *American Zoologist,* 17(2): 411–429.

1978. Brain Organization in the Cartilaginous Fishes. *In* E. S. Hodgson and R. F. Mathewson, eds., *Sensory Biology of Sharks, Skates, and Rays.* Arlington, Virginia: Office of Naval Research, pp. 117–193.

Olsen, A. M.
1984. Synopsis of Biological Data on the School Shark, *Galeorhinus australis* (Macleay 1881). *FAO Fisheries Synopsis* No. 139. Rome: Food and Agriculture Organization of the United Nations, 42 pp.

Otwell, W. S., F. J. Lawlor III, J. A. Fisher, G. H. Burgess, Jr., F. J. Prochaska, and J. M. Stevely, compilers.
1985. Manual on Shark Fishing. *Sea Grant Report No. 73.* Gainesville, Florida: Florida Sea Grant College, 44 pp.

Parin, N. V.
1966. Data on the Biology and Distribution of the Pelagic Sharks *Euprotomicrus bispinatus* and *Isistius brasiliensis* (Squalidae, Pisces). *Trudy Instituta Okeanologii,* vol. 73, pp. 163–184.

Parsons, G. R.
1985. Growth and Age Estimation of the Alantic Sharpnose Shark, *Rhizoprionodon terraenovae:* A Comparison of Techniques. *Copeia,* no. 1, pp. 80–85.

Pratt, H. L., Jr.
1979. Reproduction in the Blue Shark, *Prionace glauca. Fishery Bulletin,* 77(2): 445–470.

Pratt, H. L., Jr., and J. G. Casey
1983. Age and Growth of the Shortfin Mako, *Isurus oxyrinchus,* Using Four Methods. *Canadian Journal of Fisheries and Aquatic Sciences,* vol. 40, pp. 1944–1957.

Purdy, R. W.
1984. Appendix—A Key to the Common Genera of Neogene Shark Teeth. *In* L. W. Ward and K. Krafft, eds., *Stratigraphy and paleontology of the outcropping Tertiary beds in the Pamunkey River region, central Virginia Coastal Plain-Guidebook for Atlantic Coastal Plain Geological Association 1984 field trip.* Atlantic Coastal Plain Geological Association, pp. 210–215.

Ramfjord, S. P., and M. M. Ash, Jr.
1971. *Occlusion.* 2d ed. Philadelphia: W.B. Saunders and Co., 427 pp.

Randall, J. E.
1961. Let a Sleeping Shark Lie. *Sea Frontiers,* 7(3): 153–159.

1973. Size of the Great White Shark (*Carcharodon*). *Science,* 181(4095): 169–170.

1977. Contribution to the Biology of the Whitetip Reef Shark (*Triaenodon obesus*). *Pacific Science,* 31(2): 143–164.

1987. Refutation of Lengths of 11.3, 9.0, and 6.4 m Attributed to the White Shark, *Carcharodon carcharias. California Fish and Game,* 73(3): 163–168.

Randall, J. E., and G. S. Helfman
1973. Attacks on Humans by the Blacktip Reef Shark (*Carcharhinus melanopterus*). *Pacific Science,* 27(3): 226–238.

Randall, J. E., and M. F. Levy

1976. A Near-Fatal Shark Attack by a Mako in the Northern Red Sea. *Israel Journal of Zoology,* vol. 25, pp. 61–70.

Ripley, W. E.

1946. The Soupfin Shark and the Fishery. *California Department of Fish and Game, Fishery Bulletin* 64, pp. 7–37.

Robeco Chemicals, Inc.

1984. Robane-Supraene-Robuoy-Robecote. Robeco Chemicals, Inc., 99 Park Ave., New York, NY 10016 [privately printed], 23 pp.

Ronsivalli, L. J.

1978. Sharks and Their Utilization. *Marine Fisheries Review,* 40(2): 1–13.

Rosenthal, M. L.

1984. Biological Role and Practical Uses of Squalene and Squalane. *In* I. Lubowe and F. V. Wells, *Cosmetics and the Skin.* New York: Reinhold Publishing Corp., 646–653.

Russo, R.

1984. Whitetip—The Cave Shark. *Sea Frontiers,* 30(1): 30–36.

Ryan, P. R.

1981. Electroreception in Blue Sharks. *Oceanus,* 24(4): 42–44.

Schaeffer, B.

1967. Comments on Elasmobranch Evolution. *In* P. W. Gilbert, R. F. Mathewson, and D. P. Rall, eds. *Sharks, Skates, and Rays.* Baltimore: Johns Hopkins University Press, pp. 3–35.

Schaeffer, B., and M. Williams

1977. Relationships of Fossil and Living Elasmobranchs. *American Zoologist,* 17(2): 293–302.

Scharold, J., and F. G. Carey

1985. Vertical Movements and Tailbeat Frequencies of Free-Swimming Blue Sharks. *American Society of Ichthyologists and Herpetologists, Program and Abstracts, 65th Annual Meeting,* 9–14 June, University of Tennessee, Knoxville.

1986. Behavior of the Free-Swimming Blue Shark: Depth and Speed. *American Society of Ichthyologists and Herpetologists, 66th Annual Meeting, Program and Abstracts, Second Annual Meeting American Elasmobranch Society,* 15–21 June, Victoria, British Columbia.

Scholander, P. F.

1957. The Wonderful Net. *Scientific American,* 196(4): 96–107.

Schwartz, F. J.

1983. Shark Ageing Methods and Age Estimation of Scalloped Hammerhead, *Sphyrna lewini,* and Dusky, *Carcharhinus obscurus,* Sharks Based on Vertebral Ring Counts. Proceedings of the International Workshop on Age Determination of Oceanic Pelagic Fishes: Tunas, Billfishes and Sharks, February 15–18, 1982. *U.S. Dept. of Commerce, NOAA Technical Report 8,* pp. 167–174.

1984. Occurrence, Abundance, and Biology of the Blacknose Shark, *Carcharhinus acronotus* in North Carolina. *Northeast Gulf Science,* 7(1): 29–47.

Seigel, J. A.
1978. Revision of the Dalatiid Shark Genus *Squaliolus:* Anatomy, Systematics, Ecology. *Copeia,* no. 4, pp. 602–614.

Seymour, H. H. and K. S. Danberg
n.d. Shark. *Sea Grant Advisory Service Bulletin.* Newark, Delaware: University of Delaware, 6 pp.

Smeets, W. J. A. J., R. Nieuwenhuys, and B. L. Roberts
1983. *The Central Nervous System of Cartilaginous Fishes: Structural and Functional Correlations.* Berlin, Heidelberg, New York: Springer-Verlag, 266 pp.

Snelson, F. F., Jr., T. J. Mulligan, and S. E. Williams
1984. Food Habits, Occurrence, and Population Structure of the Bull Shark, *Carcharhinus leucas,* in Florida Coastal Lagoons. *Bulletin of Marine Science,* 34(1): 71–80.

Snodgrass, J. M., and P. W. Gilbert
1967. A Shark-Bite Meter. *In* P. W. Gilbert, R. W. Mathewson, and D. P. Rall, eds., *Sharks, Skates, and Rays.* Baltimore: Johns Hopkins University Press, pp. 331–337.

Springer, S.
1950. Natural History Notes on the Lemon Shark, *Negaprion brevirostris. Texas Journal of Science,* 2(3): 349–359.

1960. Natural History of the Sandbar Shark, *Eulamia milberti. Fish and Wildlife, Fishery Bulletin* 178, vol. 61, 38 pp.

1963. Field Observations on Large Sharks of the Florida-Caribbean Region. *In* P. W. Gilbert, ed., *Sharks and Survival.* Boston: D.C. Heath Co., pp. 95–113.

1979. A Revision of the Catsharks, Family Scyliorhinidae. *U.S. Dept. of Commerce, NOAA Technical Report NMFS Circ. 422,* 152 pp.

Springer, S., and H. R. Bullis, Jr.
1960. A New Species of Sawshark, *Pristiophorus schroederi,* from the Bahamas. *Bulletin of Marine Science of the Gulf and Caribbean,* 10(2): 241–254.

Springer, S., and P. W. Gilbert
1976. The Basking Shark, *Cetorhinus maximus,* from Florida and California, with Comments on its Biology and Systematics. *Copeia,* no. 1, 1976, pp. 47–54.

Springer, S., and G. H. Burgess
1985. Two new Dwarf Dogsharks (*Etmopterus,* Squalidae), Found off the Caribbean Coast of Colombia. *Copeia,* no. 3, pp. 584–591.

Springer, V. G.
1964. A Revision of the Carcharhinid Shark Genera *Scoliodon, Loxodon,* and *Rhizoprionodon. Proceedings of the U.S. National Museum,* 115(3493): 559–632.

1982. Pacific Plate Biogeography with Special Reference to Shorefishes. *Smithsonian Contributions to Zoology,* 367, 182 pp.

Squire, J. L., Jr.
1967. Observations of Basking Sharks and Great White Sharks in Monterey Bay,

1948–50. *Copeia,* no. 1, pp. 247–250.

Stevens, J. D.
1983. Observations on Reproduction in the Shortfin Mako, *Isurus oxyrinchus. Copeia,* no. 1, pp. 126–130.

Stevens, J. D., and J. R. Paxton
1985. A New Record of the Goblin Shark, *Mitsukurina owstoni* (Family Mitsukurinidae), from Eastern Australia. *Proceedings of the Linnean Society of New South Wales,* 108(1): 37–45.

Stillwell, C. E., and N. E. Kohler
1982. Food, Feeding Habits, and Estimates of Daily Ration of the Shortfin Mako (*Isurus oxyrinchus*) in the Northwest Atlantic. *Canadian Journal of Fisheries and Aquatic Sciences,* 39(3): 407–414.

Strasburg, D. W.
1958. Distribution, Abundance and Habits of Pelagic Sharks in the Central Pacific Ocean. *U.S. Fish and Wildlife Service, Fishery Bulletin* 58, pp. 335–361.

Taniuchi, T., N. Kuroda, and Y. Nose
1983. Age, Growth, Reproduction, and Food Habits of the Star-Spotted Dogfish *Mustelus manazo* Collected from Choshi. *Bulletin of the Japanese Society of Scientific Fisheries,* 49(9): 1325–1334.

Taylor, L. R.
1985. White Sharks in Hawaii: Historical and Contemporary Research. *Memoirs of the Southern California Academy of Science,* vol. 9, 24 May, pp. 41–48.

Taylor, L. R., L. J. V. Compagno, and P. J. Struhsaker
1983. Megamouth—A New Species, Genus, and Family of Lamnoid Shark (*Megachasma pelagios,* family Megaschasmidae) from the Hawaiian Islands. *Proceedings of the California Academy of Sciences,* 43(8): 87–110.

Templeman, W.
1984. Migrations of Spiny Dogfish, *Squalus acanthias,* and Recapture Success from Tagging in the Newfoundland Area, 1963–65. *J. Northwest Atlantic Fishery Science,* 5(1): 47–53.

Tester, A. L.
1963. The Role of Olfaction in Shark Predation. *Pacific Science,* vol. 17, pp. 145–170.

Thomas, M. C.
1968. *Fossil Vertebrates—Beach and Bank Collection for Amateurs.* Hendersonville, North Carolina: M.C. Thomas, 72 pp.

Thorson, K. S., and D. E. Simanek
1977. Body Form and Locomotion in Sharks. *American Zoologist,* 17(2): 343–354.

Thorson, T. B.
1971. Movement of Bull Sharks, *Carcharhinus leucas,* between Caribbean Sea and Lake Nicaragua Demonstrated by Tagging. *Copeia,* no. 2, pp. 336–338.

1972. The Status of the Bull Shark, *Carcharhinus leucas,* in the Amazon River. *Copeia,* no. 3, pp. 601–605.

1976. The Status of the Lake Nicaragua Shark: An Updated Appraisal. *In* T. Thorson, ed., *Investigations of the Ichthyofauna of Nicaraguan Lakes.* University of Nebraska, Lincoln: School of Biological Sciences, pp. 561–574.

Thorson, T. B., D. E. Watson, and C. M. Cowan
1966. The Status of the Freshwater Shark of Lake Nicaragua. *Copeia,* no. 3, pp. 385–402.

Thorson, T. B., and E. J. Lacy, Jr.
1982. Age, Growth Rate and Longevity of *Carcharhinus leucas* Estimated from Tagging and Vertebral Rings. *Copeia,* no. 1, pp. 110–116.

Tricas, T. C.
1985. Feeding Ethology of the White Shark, *Carcharodon carcharias. Memoirs of the Southern California Academy of Sciences,* vol. 9, pp. 81–91.

Tricas, T. C., and J. E. McCosker
1984. Predatory Behavior of the White Shark (*Carcharodon carcharias*), with Notes on its Biology. *Proceedings of the California Academy of Sciences,* 43(14): 221–238.

Uyeno, T., K. Nakamura, and S. Mikami
1976. On the Body Coloration and an Abnormal Specimen of the Goblin Shark, *Mitsukurina owstoni* Jordan. *Bulletin of the Kanagawa Prefectural Museum (Natural Science),* no. 9, pp. 67–70, plates 1 and 2.

Webb, P. W.
1984. Form and Function in Fish Swimming. *Scientific American,* 251(1): 72–82.

Weihs, D., R. S. Keyes, and D. M. Stalls
1981. Voluntary Swimming Speeds of Two Species of Large Carcharhinid Sharks. *Copeia,* no. 1, pp. 219–222.

Wilson, E. O.
1985. Special Report: In Praise of Sharks. *Discover,* 6(7): 40–42, 48–53.

Wolfson, F. H.
1986. Occurrences of the Whale Shark, *Rhincodon typus* Smith. *In* T. Uyeno et al., eds., *Indo-Pacific Fish Biology: Proceedings of the Second International Conference on Indo-Pacific Fishes.* Ichthyological Society of Japan, Tokyo, pp. 208–226.

Wood, L.
1986. Megamouth: New Species of Shark. *Sea Frontiers/Sea Secrets,* 32(3): 192–198.

Wourms, J. P.
1977. Reproduction and Development in Chondrichthyan Fishes. *American Zoologist,* 17(2): 379–410.

Index

Numbers in boldface indicate illustrations or plates.

N

O

P

R

S

T